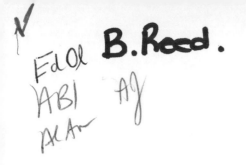

Ed Ol B. Reed.
ABI AJ
AcAr

A ROMANY IN THE COUNTRY

A ROMANY IN THE COUNTRY

By
ROMANY
G. BRAMWELL EVENS

ISIS

LARGE PRINT

Oxford

First published in Great Britain 1932
by The Epworth Press

Published in Large Print 2002 by ISIS Publishing Ltd.
7 Centremead, Osney Mead, Oxford OX2 0ES
by arrangement with Mrs Romany Watt.

British Library Cataloguing in Publication Data
Evens, George Bramwell
A Romany in the country. –
Large print ed. – (ISIS reminiscence series)
1. Natural history – Great Britain
2. Large type books
I. Title
508.4′1

ISBN 0–7531–9742–1 (hb)
ISBN 0–7531–9743–X (pb)

Printed and bound by Antony Rowe, Chippenham

To J. C. Beattie,
MY HEADMASTER

A ROMANY IN THE COUNTRY

in which you are re-introduced to:

RAQ .... *My Spaniel*
JOHN FELL.................. *Gamekeeper*
JERRY .... *Poacher*
ALAN AND JOE .... *Farmers*
NED .... *Village Postman*
JOHN RUBB .... *Angler*

In publishing this series of Nature Talks I have purposely omitted much of the village gossip which is to be found in my two previous books, and in consequence, one or two village characters are omitted. This has been done that more space might be devoted to Nature lore and fact.

August, 1932   G.B.E.

CONTENTS

INTRODUCTION

I do not remember the time when I did not take a keen interest in all wild life. A grass field, a hedge with a linnet swaying on the branch of a wild rose has always meant heart's delight to me.

When, however, I began to study birds and animals seriously, I then felt the need of living where they lived. Hence a caravan became mine, and a tent. There are many kinds of tents, and I have had experience of them all. I have stowed myself away in a small Scout's tent, where my toes had a perfect genius for finding the daylight, and where Raq, my spaniel, had to nestle up against me, and thus provided a hot-water bottle that never grew cold. I know what a comfortable place the gipsy tent of my ancestors was, with its blankets stretched tightly over arched hazel poles. But not one holds my affection as the bell-tent which I have used for many years.

It is dyed brown, and the light filtering through gives everything a warm tone, far better than those green affairs which make every one look sickly, or suffering from pernicious anaemia.

All I know of its ancestry is that John Rubb procured it for me. I experimented with many, and finally chose it to be my own, for it manfully withstood the shock of Scotch gales and the moist inquisitiveness of Scotch mist.

How it came into his shop I cannot say, but on it

in dark paint stands the number 5. No doubt it has known the rigours of army life — it may even have been under fire. Now it spends its mature age quite peacefully. In winter it hangs from a beam in a barn, and there are evidences that its only attackers are bright-eyed field-mice. Fortunately they have only found its "flounce" palatable, and the holes aid ventilation.

It is a joyous moment, both for me and the tent, when we unfold its long skirts to the summer sun.

"At last," I hear it say, as I ram home the tent-pole in its crown. "At last," I say aloud with triumph, as the pole stands erect, all my Romany blood tingling with ecstasy and expectation.

I always place the entrance, if I can possibly manage it, so that it greets the morning sun. Then I stretch its skirts so that they fall just outside my wooden floor-boards.

Soon the sound of the mallet is heard driving in the pegs, the guy-ropes are pulled taut, the flounce is hooked — and there we are!

Those guy-ropes — what a nuisance they can be — a trap to the unwary and a sure damager of tender toes! How many times during the night watches have I emerged from the tent in pouring rain in order to slacken them, lest the strain of contraction should bring disaster! Now, however, I have found out a "wheeze" which prevents such a midnight excursion on wet grass: I stand the pole upon a small stone. Then, when I hear the guy-ropes straining, I slide the pole off the stone, and thus reduce the tent's height by a couple of inches.

But when I am in bed those ropes often tell me that I have visitors. Once a young fox blundered against them in the early hours of the morning, and was mightily surprised when my tousled head looked from under the flounce at him.

Once the whole tent shook even as a web shakes with a fly's struggles, and it was my turn to be surprised, for with staring eyes and snuffing muzzle a cow pushed her head through the open entrance. A moment later Raq reminded her that it was indelicate for her to intrude, and she moved off, leaving half a dozen of my guy-ropes in a very sickly condition.

In the open entrance of the tent or on the steps of the caravan I like to lounge. Sheer lazy restfulness are these hours — Timelessness at its best.

Here I often muse on the riches that are mine — the treasures of the open air and the wealth in my friends, especially those whom I meet in the lanes and fields. Each one has a different and engaging personality. There are Alan and Joe, large-hearted, jolly, generous, hospitable farmers, who are always ready to initiate me into the art of dealing with live stock, and never tire of answering my questions. Their home on the fells is my home, and when tired of city life and its din, I can always find a retreat there and listen to the music of river and beck, whilst the curlew with his pipe takes the place of Pan.

Then there are other comrades of the soil. What a character Ned, the old postman, is. He is something of a philosopher, more of a poet, and simple enough to show his enjoyment of using big words. His whimsical outlook on birds and flowers gives me hours of pleasure.

No one could ever meet Jerry without liking him. He is a born poacher, it is true, but never have I known him take a rabbit or hare for private gain. The villagers almost idolize him, and the aged and sick have received countless delicacies from his capacious pockets. Even John Fell, the gamekeeper, likes him in spite of his delinquencies, and Jerry respects John and the estate he has to guard. None of these men have the polish of the schools, but they are graduates in the lore of the countryside, and into their wells of knowledge it is my joy to dip deep.

Sometimes, too, I sit and marvel at my own inconsistencies. There was a time when I was keen on using a gun. Those days are past and now I only "shoot" things with my camera. Yet I can still enjoy a day with the rod, especially when John Rubb is my companion. He it is who taught me the difficult art of casting the fly — he it is who still points out to me the treasures of the river.

So as I sit in my tent I see the faces of my friends and hear them say, "Silver and gold have I none, but such as I have give I unto thee." I am indeed rich in my friends.

Just beyond the wicket-gate I can see a hayfield, now standing bare — even the rakings have been safely gathered. Some time ago all my family tried its hand at haymaking. It was a hot day, and I had appointed myself as general manager, over-seer, superintendent, Mussolini! but, to my sorrow, I found that my staff were more anxious to see me use a rake than to be told how to do it! Some of them had seized a pitchfork, and it had been as hard work to dodge the prongs of those amateurs as it had been to make the cocks!

Then the farmer had come down and surveyed our work. He said nothing, but, taking a rake, by a few deft touches he transformed our dishevelled-looking mounds into fair imitations of those gathered by his wife and himself.

"Ye want to comb 'em oot a bit," he had said, trimming each cock downwards, "so as t' watter 'll run off."

We all looked on and saw the difference between the amateur's and the professional's touch. How easily he plied the rake! When we used it, its prongs had tried to lift up a dozen sods along with the hay. But it never faltered in his capable hands. We were only fit to handle the fork if each one had a field to himself, but in his hands its poise and balance, and the deftness with which he speared up half a cock, was a thing to marvel at — thrust and lift were combined in one movement.

"I hope we haven't spoilt any of it," my wife had said half-fearfully.

"Na," he had answered, "ye've not spoilt owt."

But I noticed that he never said that we had been of much help.

What a lot of meaning country folk can put into things which they never say!

I like the tent best, however, in the evening. Like a white moth it sways against the deep shadow of the pines — not underneath them, for there it would catch all the rain-droppings, and I should be pestered with midges. As it is various insects think they have a right of entry. This year the daddy-long-legs have made it a rendezvous. Two years ago there were earwigs everywhere. After I have lit my lamp and its warm glow flows out of the entrance on the dew-laden grass, then the moths — some white, others aglow in amber, many draped in soft brown — come and flutter round its brightness. So long as they do not flutter on my face, I do not object. I must say I do not like one exploring the cavities of my ear just when I am dropping off to sleep!

Tent-life has its ups and its downs, but there are more ups. I love that hour when everything is hushed outside and Nature seems to be holding her breath lest she should wake her roosting children.

Then I light my lamp, prop myself up with my pillows, draw out my pipe, and read.

Beside me lies Raq on a sack. He is usually sound asleep, and sometimes his body quivers and he emits tiny barks, which sound as though they are echoing from distant lands, where he is chasing rabbits which he always catches.

"Steady on, old man — turn your engine off," I say, if he grows too vociferous, and at the sound of my voice he looks up wonderingly, licks my hand apologetically, gives a lazy wag, and is off hunting again.

Then the tent gently heaves to the passing breeze, and I look out on a world of half-shadow and deep shadow. Orion and the Pleiades and Cassiopeia wink at me through the open flap. A fragrance, just a suggestion of angel's breath, steals in. I put out my pipe, for the soul of the honeysuckle is paying me a visit, and behind it all I am conscious of a Presence who "calleth them all by name" — Saturn, Jupiter, Sirius — mighty orbs — and yet has not overlooked touching a wild flower with the fragrance of Paradise.

CHAPTER ONE

Stalking a Deer

Passing the fringe of a wood I was surprised to notice Raq stop, "point," and wag his tail. A moment later Jerry stepped out.

"Ye're jist the very feller. There's a deer gone by."

"A deer, Jerry?" I said eagerly. "Couldn't we stalk her?"

"I fancy she's in the Redwood cover," said he. "I come across 'er 'slots' on t' groond and they were quite fresh ten minutes sin. If she's in there, she'll 'ave to cross a lot o' oppen country afore she can find another restin' place and she won't fancy all t' countryside seein' her. But we'll 'ave to mind and enter Redwood from t'other side or t' wind'll tell 'er we're there, specially wi' Raq at yer 'eels."

"Was it a red deer?" I asked, as we left the road and took to the fields. He shook his head. "There are none o' them kind i' these parts. Up i' Scotland ye see 'em, and t' gentry go up there to shoot 'em. I reckon if we git a squint on't (and if we git that we're lucky) we shall find it's a roebuck."

"Something like the fallow deer I've seen in gentlemen's parks?"

1

"Aye, aboot same size, but not so light in colour, and they've none of them yeller spots on their flanks. T' roebucks are much darker i' colour — red brown, wi' a black muzzle — they're much 'ardier an' all."

"Aren't the fallow-deer hardy then?"

"They'd soon die oot if they warn't protected i' parks. But them roebucks thrive in t' woods. They're not much good for eating, and mebbe that's one o' t' reasons why there's a good many on 'em still knockin' aboot. When a thing 'asn't much taste, then it 'as more chance o' survivin'."

"We'll 'ave to cross t' river," said my companion. "Ye don't mind a bit o' a wetting'?"

"Not if it means the sight of a deer."

We plunged down towards the river edge, and came to a tree growing almost at right angles to the cliff face.

"Crawl after me along t' trunk, and then slide doon you thick branch. There's a ridge o' rock waitin' fer ye. I'll go first, fer I could cross it wi' me eyes shut. I don't reckon there's many i' these parts as knows it."

Jerry was soon standing on the ridge, and I saw that the water was almost up to his knees. "Tak' yer time," he urged, "and when ye start to slither doon I'll fix yer feet in t' proper foothold. It's fairly deep each side o' t' rock."

As I was swarming down the friendly branch I heard a huge splash. "It's only Raq," shouted Jerry, laughing. "He warn't waitin' 'is turn fer tree trunks.

2

'E took a header, and t' current's washin' 'im doon stream, but e's all right."

I soon felt myself on the ridge of rock, and as we started through the water for the opposite bank Jerry said, "Don't lift yer feet more'n ye can 'elp. It upsets yer balance. Slide 'em along one at a time i' front o' ye, and throw t' weight o' yer body forrard — feel yer way, ye know."

When we reached the far side Raq was there to welcome us, and as we began our détour, so as to enter Redwood against the breeze, Jerry said, "A wettin' never 'arms a body so long as ye keep movin' afterwards. It's sittin' aboot, and 'anging roond doin' nowt, as gives folks colds."

Thereafter, at Jerry's command, we trudged on in silence, broken only by the sluicing of the water in our boots.

As we neared the wood Jerry sat down and emptied his boots. I followed suit, so that we could walk noiselessly into the wood.

"I reckon I know just aboot where she'll be," Jerry said, and he led the way, taking advantage of every bit of cover. Sometimes we had to crawl, but we never came to an open space without my companion first scrutinizing every corner of it before we showed ourselves.

Once, with great satisfaction, he pointed to a track on the soft ground. It looked to me like the footprint left by a sheep, though with a more delicate outline and more pointed at the toes.

Once, as we were creeping by the side of some undergrowth, a magpie, in the distance, gave its rattling alarm. Jerry made a wry face. "If 'e once catches sight on us, we're done," he whispered. "We'll lie low a bit. It's mebbe just chance that 'as made him sing oot." As we crouched down, the dog cuddled up and gave me a questioning glance, as though saying, "What's the game, master?" I put my hand on his head, and he lay quiet.

Hearing no more alarms from the magpie, we once again set off, though my back was beginning to feel the strain of spending so much time on all-fours.

When we reached what I judged to be the centre of the wood, Jerry motioned to me to keep still. He went on alone, and his movements reminded me of a Red Indian. He writhed and crawled under bushes, pushing back and intertwining those that blocked his path and might have caused the least sound.

Then, with delight on his face, he turned round and beckoned me to come on. I peered through the screen of stems, and at first saw only a glade warmed by withered bracken. The silence was profound, heightened only by the silver notes of a great tit, and the distant "pink-pink" of a chaffinch. But I saw no deer.

Jerry slowly raised his hand and pointed steadily at what looked like a smudge of sombre shadow. Still I could see nothing. A moment later there was a slight movement, the flicker of something less dark than the shadow itself. Then I picked out a dainty head, with its muzzle lying between its forelegs. I squeezed Jerry's arm as a signal that I had seen the deer. The flicker must have been the movement of its ears.

For some moments we both sat rigid, holding the dog between us. Jerry touched my arm, pointed to his own ears, and then jerked his thumb in the direction of the deer.

As she lay there, inert and passive, hardly distinguishable from soiled bracken or shadow, I could see that she was keenly alert to every sound.

Sometimes her ears pointed forwards, both focussed on one particular place. Then, turning on well-oiled pivots, one wheeled left, and the other right. Then, never remaining stationary for more than a moment, like large sea-shells they turned backwards to catch the murmur of whispering branches, snapping twigs, moving birds, and the slight, abrupt, fateful unknown sound which is none of these, and yet must be interpreted.

For a time her beautiful eyes were veiled by lids fringed with long lashes. Then she raised her head, and we looked for a moment into slumbrous pools of innocence and inquisitiveness — lustrous, languorous.

What happened next I can only surmise, but I fancy that in trying to rest my legs, which were aching with my cramped position, I must have put one of my boots down on Raq's foot, for he jumped up.

5

How quickly the doe interpreted the movement must be seen to be appreciated. As quick as thought, she was up and away. Her action was a combination of a loosened spring and a quick bullet. All we saw was the flash of light-coloured hindquarters, and all we heard was a dainty step on fallen leaves — then silence.

As I stretched myself slowly, Jerry said, "Stiff, are ye?"

"But it was well worth it, Jerry," said I.

"'Aven't ye niver seen one afore?" he asked, as we retraced our steps.

"Never within forty yards. What ears! I have never seen anything more sensitive or more responsive in my life."

"If your'n were as sharp as 'er's, ye'd be ravin' mad i' a week. Did ye think, when ye were watchin' 'er, 'ow many things she 'eerd as we didn't? That lil' doe could pick up the sigh of a tom-tit, I reckon — and if your ears were like 'er's the big row of a city 'ud drive ye crazy."

For the remainder of our walk we were silent. Raq took advantage of our preoccupation to scuttle a rabbit or two from their cosy seats. The flopping of his big ears, as he chased one to its burrow, recalled me to earth, and I said,

"Raq's ears are not much use to him. He has to raise them a little when he listens intently."

"That's partly through not usin' 'em, and partly due to t' dog-breeders. I've 'eard that all wild dogs

6

used to 'ave ears summat like collies, standin' up straight, and then a slight forward flap. 'E'd either to use 'is ears when 'untin', or starve. Then when 'e took up wi' 'uman bein's 'e grew lazy, and didn't use 'em so much, and so t' muscles grew slack i' course o' time. Then t' dog-fancier did t' rest by interbreedin' slack-eared 'uns with more slack-eared 'uns, until ye get 'em wi' big wallopers like Raq's."

When we had emerged from the wood Jerry pointed to what looked like a clod of earth lying in a field some distance away.

"That's an owd hare. Git yer field-glasses on 'er and tell me 'ow she's lyin'."

"Quite squat," I replied, "with her head turned towards us."

"An' 'er ears?"

"Pressed back."

"Then she's seen us," said he. "Them ears of 'er's are set to catch any danger comin' from behind, while 'er eyes keep a lookout in front. If ye want to spend an interestin' day some time keep yer eyes oppen fer all the different kinds o' ears ye come across. Animals as is 'unted can turn 'em backards or forrards or sideways. Them 'as 'unts, 'as their ears allus pointing forrards."

"What about birds, then? They've no ears one can see, and yet their hearing is very keen."

"No ears outside, ye mean. 'Ow would ye like to see Raq's big floppers stuck on that feller," said he, pointing to a thrush perched above us.

"His ears," I said laughing, "would always be catching in the branches."

"Come on, long-lugs," Jerry called to Raq, "we're goin' 'ome."

CHAPTER
TWO

A Shepherd and
His Lambs

Raq and I had started out fairly early, and at last found ourselves high up on the fells which overlook the farm.

There had been several days' frost, but the morning was bright, and the hills looked at their best. Every crest was crowned with fine snow. Here and there patches of greener moss and grass peeped out. The gullies lay deep in blue shadow.

Standing on the side of a stiff ascent, I paused to enjoy the solitude. From the valley below came the faint but distinct sounds of farm life. The crowing of cocks and the quackings of ducks sounded quite musical when muffled by the intervening miles. Now and then one caught the tinkle of childish voices or the impatient shout of a farmer at his stock.

Then I heard a shrill whistle, and a moment later I saw a flock of sheep running down from the heights opposite. Against the snow they looked soiled, but against the grass they looked like so many faltering drops of a white cascade. How sure-footed they were on those treacherous rocks and slopes!

Further on, and on my side of the fell, I saw the dark figure of a shepherd. His commands were meant for a dog which I could not at first see, but which understood the piping morse code of his master.

Later, a small black-and-white lanky collie came over the crest, some distance behind the sheep which he had rounded up. He soon caught up with them, but immediately the whistle sounded he sank to earth, and waited until his master gave the signal for him to "bring them on."

In a few moments the flock were on lower ground, and another whistle brought the dog across to his master's side, where he lay and looked up at him as much as to say, "They're all there, aren't they, master?"

"That's a knowing dog of yours," I said by way of introduction, at the same time offering him a smoke.

"Aye, he's a good dog and saves me no end o' walkin'."

I looked over the long chain of hills separated by deep valleys. His district extended as far as I could see.

"A small aeroplane would suit you, I think," I said. "Then you could land on each crest and save the dip into the glens."

He smiled. "Couldn't do withoot the dog even then. Some sheep 'ave to be rooted oot fra all sorts o' queer places. Even then the dog has to master 'em, and they knows when a dog'll stand no nonsense. 'E's got to know his job and be firm wi' 'em, or else they'll double back, and give 'im no end o' trouble. Aye, firm and not 'arsh, that's what a dog's gotten to be."

"And how far do you think you tramp in a day?" I asked.

The shepherd pushed up his cap and scratched his head. "I should say nigh on to twenty miles a day, reg'lar — mebbe more, mebbe less."

"Eight hours a day?" I hazarded, with a smile.

"It'll work oot more like ten or twelve," he said simply.

"And how many sheep do you look after?"

"Thirteen hundred and fifty-seven," he answered promptly.

"And would you choose a shepherd's life if you had your time over again?"

He did not answer for a moment. "I shouldna like an indoor job," was all that he said.

Twenty miles a day in all weathers, lambing-time with its seventeen and eighteen hours per day of responsibility, thirteen hundred sheep to look after!

Bidding him good-morning, Raq and I wended our way down to the farm where Alan and Joe, my beloved friends, live.

From the hill-top we overlooked a valley which had the spirit of "leisureliness" brooding upon it, and I knew that for a few hours at least I was to be free from the tyranny of the clock. On those undulating pastures nothing hurries. The blue smoke curls up from the chimneys of the homesteads. Sometimes the fragrant smell of burning wood drifts upwards. Then, for a second, one catches the nutty incense of peat.

Already the rooks were lumbering along to their feeding-grounds, taking the same air route which had been their right of way for hundreds of generations. From the woods came the challenging, raucous call of the cock pheasant.

At the farm itself we went through the customary rites of initiation. I have never arrived without Hannah and Charlotte being firmly convinced that I was ready to die of starvation.

After fortifying ourselves, Raq and I accompanied Alan and Joe as they carried supplies to their men working in the fields.

On all sides the pastures were dotted with sheep. The "Blackfaced" were feeding on the turnips. Higher up were the Cheviots. But those which interested me most were the ewes, every one of which had a red mark on its flank.

"You haven't any Herdwicks, have you?" I asked. Joe shook his head.

"They don't thrive on this land. They need —" Here he waved his hand towards the hills now veiled in the pearly distance.

"I've seen them on the fells," I said; "small bodies, long fleece, smoky-grey, with a kind of top-knot on their head and broad bushy tails. Curly horns, too," I added.

"Only the 'tups' have horns as a rule," corrected Joe. "Did ye notice how they were marked?"

I shook my head.

"Some o' the shepherds," continued Joe, "put tar and paint on 'em like we do. But many of 'em clip their ears in different ways."

"What for?" I asked with interest.

"Oh, well," said Joe, "nothin' would hold them sheep together if they once made up their minds to flit. They can 'loup' any dyke, and some on 'em might stray in mist or fog."

"Aye," said Alan, "that's true enough, but Herdwicks as a rule keep to their own land, fences or no fences. They've bin there fer scores o' years, and —"

"Nice and tough they'd be at that age," said Joe, winking at me.

"I think I must have had several shoulders," I said, with painful memories.

"You know what I mean," said Alan, making a dive at his brother, "them and their forbears have been there ages, and if ye tak' 'em down to the lowlands, they've got a homin' instinct and will mak' their way back cross-country agen to their old heath."

"That's true," said Joe, "but if they do get lost, then them ear-cuts can't be altered or rubbed oot, and so they tell where they come from."

"And supposin' I were to tell ye that we had some sheep on oor farm which were 'key-bitted or click-forked,' or supposin' I described one as 'under key, bitted near, ritted far, a stroke down near the buttock, and one down far shoulder,' what ud ye think?" asked Alan, with a smile.

"He'd think ye'd gone daft," said Joe, "or that ye'd got the gift o' tongues."

"Well," said Alan, unperturbed, "that's how the shepherds ovver yonder describe the different ways the ears are clipped. It sounds a bit foreign-like to a townsman, I reckon."

Joe left us to go and strengthen a few shelters. These were always in readiness as hospital wards for any of his flock that might need special attention. "I'll be back soon," he called, as he strode away.

For a time I watched the lambs.

As they played, so the patient ewes browsed, constantly glancing up at the romping groups and sending out motherly warnings to their families not to get too far away.

When Joe returned I asked, "And have you no enemies to contend with — no foxes or anything?"

"Not so many here," said he. "But I have had trouble wi' 'em. Farther up there" — here he waved his hand in the direction of the fells. "I've bin out at dusk and seen the sly beggar away up on the sky-line sneakin' and circlin' round a ewe wi' twin lambs. At first I took it fer a strange dog. Then I heard him bleat, and —"

14

"Just a moment," I said, interrupting him. "You heard what bleat — the fox or the sheep?"

"The fox, o' course," said Joe.

I think I must have looked very sceptical, for Joe said —

"It's true, right enough. A fox is a reg'lar ventriloquist. You ask Jerry next time ye see 'im, and he'll tell ye how he can lie in the bracken and imitate the squeal of a rabbit."

"And what does he bleat for?" I asked.

"Well," said Joe, "by dusk, most o' the ewes have settled down, and their lambs are cuddlin' up beside 'em. Up comes Reynard and scents that there's one wi' a couple o' very young lambs, mebbe only a few hours old. Them's the only kind he's fond on — young and tender. Then he oppens his mouth and bleats."

"An exact imitation?" I asked in amazement.

"Not exact," said my friend, "but near enough to make every ewe restless and stir uneasy like. Some on 'em 'll get on their feet, thinking they've lost a lamb, and then the cunning beggar gets his chance of a savoury morsel."

I could see, as Joe spoke, the whole drama as it was played — the field with the peace of twilight enfolding the sheep, soft grey shadows dotted all over it, the lingering light towards the west silhouetting the distant hedge sharply against the sky-line. Then, into the drowsy ear of twilight would echo that metallic bleat, and instantly the anxious mother of the newly-born lambs would be on her feet facing the dark shadow that circled round her and her charges. No need to tell her what to do. Instinct bade her keep herself between the brown sleuth and her helpless young. As he sneaked round and round, so she, brave mother! would revolve also, always facing the menace.

But if by any chance she became flustered, or bad tactics left a gap between her and her lambs, there would be a lightning dash by that sneaking shadow, relentless jaws would crash on tender shoulders, a piteous cry of the mother into the oncoming night, then silence save for the piping of a curlew as it sailed to the distant estuary.

As Raq and I went home later through the farmyard we saw a large motor lorry. I could hear the barking of Nell and the bleating of ewes and lambs. Then Joe appeared in the lane, driving a part of his flock, who little knew that the lorry awaited them. I like to see a flock of sheep on a country road, but with these modern methods of transport the shepherd and his panting dog will, alas! soon be a thing of the past.

Over the grass mounds trotted the lambs, lithe of limb and carefree. Separating them from their mothers they were driven through the yard and herded into the lorry.

I watched Joe, who had tended them ere they could stand, bolt it. As the lorry started I imagined he deliberately refrained from looking at them. As he passed me he just shook his head, but we said nothing.

In the distance I could hear the ewes bleating for the lambs who would never return.

Walking home I thought of the old shepherd. He would be gathering his sheep together fearing bad weather.

Once again I saw those majestic hills. I was consoled to think that no lorry will ever take the place of a sheep-dog. Whilst the hills remain there will always be a shepherd and his dog.

Thank God for precipitous and inaccessible places.

CHAPTER
THREE

How Animals
Communicate

As a heavy shower began to fall, I took refuge in a shed which stood facing the lane. As I examined its walls, hoping to find some sleeping bat, or even to see some wise old owl blinking at me, Raq rushed out, and I saw him giving a welcome to Ned.

The old postman had on a sturdy cape, under which he guarded his precious burdens.

"A nice drop o' rain," he said, as he entered; "it'll mak' the land smell good, I reckon."

As he spoke I climbed up to look at an old nest which reposed on one of the rafters.

"It's a swallow's, I think," I said, "and in good condition, too."

He nodded his head. "There'll be a pair o' birds thoosands o' miles from 'ere that are thinkin' on't and seein' it beckonin'."

I looked out on the shower-sprayed landscape and in my mind contrasted it with the sunny skies of Africa under which they were at the moment flying.

"There doesn't seem much for them to come back to," I said.

"No," said Ned, "but Natur' is getting ready fer her guests all the same. Them wild gales o' winter were an early spring-clean. They brushed oot o' t' way a lot o' rubbish — dead wood and sich like. Now the rain'll come and coax oot the wakenin' plants, and they'll hang oot their blue and yeller banners, an' that'll 'tice the insecks and flies into the oppen — aye, I reckon the table's bein' laid fer the swallers and the martins."

"And do you think this nest," I asked, pointing to the rafters, "will be used again?"

"Not a shadder o' doot aboot it," answered he decisively, "and by the very same pair o' birds which 'ad it last year — onless any accident befalls 'em on their travels," he added.

He gazed up at the nest in an abstracted fashion. "Aye," he went on slowly, "t' cock bird he'll set off first, wi' 'undreds o' others, and they'll pick up thoosands on the way 'ere. I allus think o' t' comin' o' the birds when I 'ears that hymn, 'See the mighty 'ost advancin' — that aboot describes it."

"And you think that all the cock birds travel ahead of the hens?" I asked.

"O' course," said he; "'e comes to secure the site, and then she follers on, and —"

"But how does she find her mate again?" I broke in. "With those thousands of birds that you talk about, it must be like searching for a needle in a haystack."

19

Ned smiled. "Ye fergit that they are all makin' fer a district that each on 'em knows, and many on 'em fer farms and ootbuildin's in which they were 'atched. If each bird 'as a ren-ren —"

"Rendezvous," I murmured.

"Rondyvoo," he repeated, "then it can't be difficult fer 'em to find each other, eh?"

I looked up again at the old nest with a deeper wonder. It was only a bit of mud and a few feathers. Suddenly I saw it take the shape of a potent magnet which was surely drawing a couple of birds over arid deserts, high peaks, inland lakes, and turbulent seas — a bit of mud transformed into a home.

As I walked along with him on his round, I reverted to the coming of the birds.

"Do you think the cock comes first simply with the idea of securing a site for the future nest?"

"Well," said my companion, after a pause, "I reckon he has that in the back o' his mind — but from what I've seen, I think that 'is great anxiety is to rail off, as ye may say, a space where he can find his own food and keep others from trespassin' on't. I'm not thinkin' only o' the swallers, but o' them birds which allus feeds on livin' things."

"And supposing," I persisted, "that, after all the preparations, the little hen mate meets with some accident and never appears, or supposing that after she arrives she finds herself a widow — what will happen then?"

"Well," said he, and I thought I saw a twinkle in his eyes, "there are exceptions, o' course, but speaking ginerally there ain't a deal o' grievin' done i' Nature. A widder to-night is a wife again i' a day or two, mebbe."

"That's another o' the mysteries o' Nature," Ned continued. "Birds lose their mates and some'ow or other the fact is noised abroad in a jiffy — how, nobody knows. It's like them S.O.S.es on t' wireless."

"They must be able to communicate with each other somehow, mustn't they?"

"O' course they does. Have ye ever thought 'ow when reptiles made up their minds that they'd leave the ground fer the air and the trees, an' grow wings, they needed to mak' sounds to tell each other where they were? That's why birds talk more'n animals."

We were passing a small wood as he spoke, and he stopped and took hold of my arm. "Do ye hear them needle-like chatterin's in the tops o' the pines?" he asked.

"Blue tits?" I said.

"They're allus on the chatter just to let each member o' the family know that they arn't gettin' scattered. They can't see too well up amongst all them leaves."

"So birds cultivated their voices, I suppose," I said, "so as to tell each other their whereabouts?"

The old postman nodded. "And on t'other 'and, beasts on t' ground used their noses more'n their tongues. Ye see, up there in tree-tops, noses ain't much use. The air is purer, and the ground currents don't rise so high. So birds cultivated sight and voice, whilst groundlin's went in more fer smell and 'earin'."

"So when a bird sings on the tree-tops," I said, "it is telling of its thoughts and sentiments — a thrush, for instance, up on the elm, is telling his mate on the eggs underneath, how much he thinks of her; or advertising with his voice that he is quite willing to enter into partnership with some one who likes the quality of his voice. Is that it?"

Ned did not reply for a moment. Then he said, "Well, there's summat in what ye say. That they're communicatin' one wi' another by song I've no doot i' my own mind."

"I've been watchin' that dog o' your'n," said Ned. "If ye notice, every time he comes along this way he allus sniffs at the same places. And if we could see 'em, every other dog that comes along here makes the same halts. That nose of his, I reckon, tells 'im —"

"Scent images," I suggested; "pictures that come to his mind from smells rather than from colours, forms and sounds, and —"

"That's it now," said Ned eagerly. "He could tell ye every other dog who had passed by, and if needs be he could follow the trail till their noses met. Smell is their means o' communication."

A little farther on we passed a fine old pine which had a hole about seven feet up.

"Know what's made that hole?" he asked.

"Woodpecker I should think," said I, looking not only at the hole, but at the chips which lay at the base of the tree.

"Ye're right," said Ned; "an old 'yaffle' has bin at work there reet enough. Ever heard him when he's busy?"

I nodded. "I've seen him holding on to the trunk, with his bill driving like greased lightning against the bark," I answered.

"Ye could hear him at work above a hundred yards away, I reckon?" asked Ned.

"Rather," I answered, delighted that I could tell him something about this bird. "If the wood is still, I believe I could hear him for a quarter of a mile."

"Like a strong drill — 'R-r-r-r-r — R-r-r-r,' wasn't it?" Ned persisted.

Then with a smile he said, "That old yaffle is doin' wi' his bill what the thrush is doin' wi' his song. He's tellin' his mate by sound where she can find him. Listen!"

Some distance away we heard the borer at work.

"That's a strong call if ye like," said Ned, as we marvelled at the strength which drove at the bark.

"A trunk call, I think," said I, whereat Ned clapped me on the back and laughed heartily. "That's a good 'un," he said, "a good 'un."

"Call the dog to heel," he said. "We'll go extra quietly out o' this wood, and if we've any luck we'll see a bit o' something interestin'."

When we came to the hedge that surrounds the spinney, Ned beckoned me to look through an opening with him. Very cautiously I did so.

Rabbits were feeding and sunning themselves a few yards from their burrows. Various birds were hunting for food. We looked over so carefully and slowly, making no sound and causing no disturbance, that for a moment all the wild things went on with their business.

Then a rabbit sat up and pricked up its ears. Others that lazed immediately became alert. Birds that foraged stood at attention — the whole field became tense and knew of our presence.

"That's the wireless communication o' the wild," said Ned. "One o' them rabbits or birds sensed us, and flashed the news to the others — 'ow, no one knows."

I looked carefully around, and, screened as we were, knew that what Ned said was true.

CHAPTER
FOUR

The Wounded Heron

Raq and I were out early. It is then that one sees wild Nature at its best. The birds wake up as children do, and without waiting to rub the sleep out of their eyes, or spend moments in stretching and yawning, leap into full life at once.

Most of their thoughts at this time of the year are concentrated on providing suitable places for the coming nests. We had evidence of this, for we watched some lively duels by the side of a pond. The dog had followed me, thinking that he was going to get some fun hunting out the water-hens which lurked in the bushes, but I restrained him.

As we hid behind a convenient bush, I saw a cock bird show himself on the distant bank. A moment later there came from the bush nearest to me an excited "Croot, croot."

In another moment the two birds were engaged in deadly strife. Bills and feet were used freely, and the duel ended in the bird which I had first seen, flying to the hedge, after which I saw him no more. Evidently the newcomer had been inspecting the pond in the hope of staking out a claim, not knowing that the coveted pond was no longer "to let."

A few moments later I saw both the cock and the hen swimming together underneath the shadows of the bushes. No doubt she was congratulating him on his prowess. His tenure, however, is by no means sure. Day by day other birds will visit that peaceful pond and cast longing eyes on its quietness and suitability for a nursery, and only by the strength of his bill will he hold his own. Still, possession is a mighty asset.

Wandering on, we came across Jerry on the river bank. He greeted us, and patting Raq, said: "Come and sit doon. There's allus summat goin' on where there's plenty o' watter."

Raq loves the river, too, and when he saw me sit down his joy was manifest. There might be other water-hens he could chase amongst the rushes, or the chance of his flushing a wild duck or two.

"Better keep 'im to heel," said Jerry, "fer ye never know what may be standin' aboot."

It was well that we did so, for within a few yards we could see the shimmering waters, and by the side of a still pool Jerry pointed to a motionless grey figure.

"Shall we try and get nearer?" I asked.

"Better not," said he; "them herons is t' quickest sighted birds there is — none to beat 'em."

So for a few moments we watched the silent fisher. His great wings were folded around him as though swathed in a sheet — more ghost than bird.

Suddenly the pointed beak shot out, and in a trice we saw a fish tossed in the air, caught dexterously, and immediately swallowed.

"I thought he always speared the fish," I said.

"Not allus," said the poacher. "I reckon 'e caught that one crosswise, an' e' can't swaller 'em like that. Ye saw 'ow 'e tossed it up — that were so as 'e could catch it head first — that's the way it's got to go doon 'is throat."

After standing like a statue for some time the big bird decided to come to life, and we saw him wading in the shallows.

"What big deliberate steps he takes," I commented.

Jerry nodded. "He treads doon eels sometimes." The poacher lifted his legs and imitated the heron's walk so well that I almost laughed aloud. Then he stood still with a stick under his boot.

"'Ow would ye deal wi' an eel if ye 'ad to stand on't 'as he does?" he asked, pointing to the stick.

"I shouldn't stand long enough on it to have to deal with it," I answered.

"But t' owd 'yarn' (heron) stands on't, and it doesn't get away," Jerry persisted. "If ever ye get the chance, have a look at his toes, and ye'll find the long middle un 'as edges on't, summat like the teeth on a saw. I reckon that gives 'im a grip o' the slimy slippery beggar. I can't think what else it's useful fer."

"A sort of non-skid foot for eels," I added.

"That's aboot it noo," said my friend.

I think the bird must have heard us, for it suddenly launched itself in the air, and we watched its ungainly flight and heard its metallic hoarse clank of a call.

"I reckon," said Jerry, "that when wings were given oot, yon bird put on a ready-made pair wi'oot tryin' 'em on first. I allus think they're two sizes too big fer 'is body.

To our amazement we heard a shot ring out in the distance. The roar rolled down the glen.

The dog heard it too, and pranced around as though he expected to see some running hare or rabbit. He dashed into the cover of the hedges to see if he could find the quarry which the gun, perchance, had missed.

"That shot's a long way away, old man," said Jerry, patting him affectionately.

About ten minutes later, as we walked along the river bank, we saw Raq emerging from the undergrowth dragging a big bird in his mouth. There was a look of triumph in his eyes and about his wagging tail.

I took the bird from him in horror. It was the heron, and from its right side a red trickle of blood stained its beautiful grey-white plumage. It was still warm.

I looked at Jerry, and his eyes blazed with anger. Shaking his fist in the direction of the shot, he blurted out — "The murderin' devil." The shooter would have had a hectic five minutes if Jerry could have laid his hands upon him at that moment.

"What 'arm was it doin' 'im? What good is it to 'im?" he shouted, pointing at the dead bird, and striding angrily up and down the bank while Raq gazed at him wonderingly. He pulled himself together with a great effort and said, "I've watched 'im scores o' times waitin' fer trout and pickin' up eels. I looked on 'im as a pal o' mine." He picked up the dead heron, and examined the wound in its side. "Poor bird," he said, stroking its crested head, "I reckon ye 'ad a job to fly as far as this wi' that 'ole under yer wing. I'll gie ye a decent burial anyway," and without another word, carrying the bird under his arm, he left the river and strode away in the direction of his cottage.

I hesitated what to do. I knew exactly what Jerry was feeling, and I thought of wandering off with the dog on my own. But the old poacher turned round and beckoned me to follow. "I'll be better in a minute or two," he said, with a ghost of a smile about his lips. "If I could let oot all I feel, I reckon that t' snow ud melt as far as t' North Pole."

"Hold it in, please, Jerry," I said.

He laughed outright and the fire in his eyes cooled, but the grim setting of his mouth boded ill if ever he should come across the man who shot the bird.

At the bottom of his garden he dug a trench.

"It's a good job that 'e were dead when Raq come upon 'im," said he. "Have a look at his long bill

29

afore we bury 'im. He'd a driven it, mebbe, clean through t' dog's skull."

"Is he naturally a very savage bird, then?" I asked.

Jerry shook his head. "He's like every good-tempered feller. It tak's a lot to rouse 'im, but when 'e is roused, there's death behind every stroke o' t' bill. Savage?" he asked. "Why, ord'narily he'll let a sparrer or a finch chase 'im fer miles, and keep just behind him nagging and natterin' like an angry wench, an' 'e won't even look round at t' li'l beggars, let alone turn on 'em. All 'e asks fer is room to fish in."

"It beats me how with such a long thin neck he manages to swallow such big fish whole. I can understand big eels going down easily, but how he manages to get down trout of over a pound weight I don't know."

Jerry took hold of the bill and opened it wide. "If ye're not squeamish, push yer fist doon his throat and ye'll soon find oot 'ow 'e manages it."

No sooner was my fist in the bird's gullet than I began to understand. First of all, the throat felt as though it were made of strong gutta-percha and it expanded, and as my fist went farther down, I was conscious of tremendous pressure being exerted on it.

"Well," said Jerry, smilingly, as I withdrew my hand, "I've seen a heron swaller even a flat flounder of a pound weight. How does 'e do it?"

"The neck expanded at first," I answered; "then the muscles crushed the flat fish till it was the shape of an ordinary herring."

30

"I reckon that's what 'appens," answered my friend.

And so we buried the bird in silence, and as Jerry smoothed the top of the grave he turned and shook his fist once more in the direction of the river.

We then returned and followed the course of the river looking for the little pad marks of those wild creatures that live on its margins. Here the poacher pointed out the tiny hand-like traces of the shrew; there, the bigger ones left by the rats. The heron had left behind a great three-pronged signature, and from the size of the water-hen's impressions one would have thought that it had been made by a big bird.

"What's that?" asked Jerry, pointing to a single line of tracks which led from a small wood right to the edge of the waters.

I saw the imprint of a cloven hoof and said, "Sheep."

"Not so bad," said Jerry, "but they're daintier than a sheep's, and run in a straight line. The hind foot fits into t' slot o' front foot, and they've bin made by two toes that are sharp and pointed. Have ye forgot 'ow we tracked yon deer?"

I was thrilled. I could see in my mind the grey morning mist on the river — the stillness of the trees — the quiet parting of the gorse-bushes, and then a dainty brown elf walking, lightly as a thought, towards the rippling water. What wouldn't I have given to have seen it?

CHAPTER
FIVE

How Birds Fly

As John Fell and I entered a field, a lark sprang in the air and poured out its life in care-free song. We stood and watched it as it mounted higher and higher. Suddenly the song died in its mouth, and we saw the dash of a sparrow-hawk, which seemed to have sprung from nowhere. I felt that the little singer had not the ghost of a chance of escape, and a huge dry lump came into my throat. But as the hawk swooped at its quarry, the lark seemed to furl its wings, and with a nose-dive which made me gasp with astonishment, landed right in the shelter of a protecting gorse-bush. The hawk, a fraction of a second too late, sheered away from the perilous prickles, and, catching sight of Raq nosing about the hedge-bottom, made off towards a distant wood.

"Lucky bird," said the gamekeeper, with satisfaction.

"What a thrilling dive to safety it made!" I said. "Let us wait and see whether it comes out of the bush."

I clambered on to a gate, and John, lighting his pipe, leaned up against it. We had not long to wait.

Within five minutes of the hawk's disappearance, the singer was again in the air. All thought of the raider must have been banished. No fear of the past weighted those dancing notes, no apprehension cut short its trills.

"That little 'un up there," said John, pointing with his pipe to the dark speck in the heavens, "beats all yer aeryplanes. He's up now ower three hundred feet. Gone up straight as a die. There's no machine that I knows of that'll go straight up from t' ground like yon bird."

Now he had reached the apex and began slowly to descend. Still the song poured forth, though I thought that on the downward flight the dashing triumph of upward song was missing. The solo continued until within thirty feet of the ground, then it abruptly ceased, and a quick dive brought the bird to the ground.

"Ever looked at the foot of a lark?" asked John, as we tramped on.

I shook my head.

"It's got a long hind toe — very long. Nobbudy knows why it's so long, but I've heerd it said as 'ow the length on't helps it to alight wi'oot a shock. They say it touches t' ground afore the rest o' t' foot, and acts like a spring. Mind ye, I've never bin near enough mysen to see it."

"I've many a time rushed to the place where a lark has alighted," I said, "in the nesting-time, thinking I should find its nest easily, but I've never had any luck."

John laughed. "A lark never comes doon right on its nest. That ud be givin' t' show away. A lark acts summat like a curlew. He flies doon to t' ground some distance from his home, and then, if t' coast's clear, runs along to it, dodgin' in and oot o' t' grasses. Ye gin'rally find t' nest when ye're not lookin' fer 't, I reckon."

"I suppose you can tell any bird by its flight, John?" I asked, as we wended our way towards his cottage.

"Just aboot," he answered laconically. "Finches and wagtails fly wi' a sort of loop motion — a dip and rise action, short and often, ye know. Then there are what I call lazy fliers." He pointed to a number of rooks that were making their way towards a distant rookery. "Look 'ow heavily they use their wings. They look as though they waft the air, not cut it clean like that there sparrow-hawk. If ye watch him ye'll see him clip his wings to his sides wi' a'most a click — that means speed."

I looked at the rooks through my field-glasses.

"Their flight feathers never seem to fit tightly together — they are all loose at the ends — almost ragged," I said.

"Aye," said John, "then there's the plover. He's another lazy un — looks as though it tak's 'im all his time to keep i' th' air. Flap, flap, flap, goes his wings, which seem to be two or three sizes too big fer 'im to manage. But fer all their lazy looks, them rooks and plovers are reg'lar stunt fliers."

As John spoke we heard the cry of a plover in a neighbouring field.

"Call the dog to heel," he said, "and we'll send him in."

Raq came obediently to my call, and by a little manoeuvring I managed to send him into the field which John pointed out. We stood under the shadow of the hedge and watched.

No sooner had the dog entered the field than the plover was up on the wing. For a moment or so it circled, emitting its plaintive cry, and I noticed its lumbering flight and its club-shaped wings.

But as Raq ran on, the bird suddenly changed its tactics. Gone was its lethargy and seeming clumsiness. It became a thing of lightning and quicksilver. Like a falling rocket it dived at the dog, and from where we stood we could hear the menacing "woof" of its wings as it skimmed the dog's head. He, with his nose to the ground, was startled, and looked around to see what had caused it. Even as he looked up the plover repeated his dive, and it was amusing to see Raq make for the cover of the hedge.

As the dog changed his direction, that bird acted like one possessed. Like a streak of forked lightning it sped downwards, throwing itself now to the right, now to the left, without checking its impetuous speed. It defied every law of gravity. It twisted, without tying itself in a knot; it turned by a flick of its wings, instead of using its tail — rolling, cork-screwing, nose-diving. When last we saw it it was looping-the-loop.

"What do ye think on't?" asked John, his voice ringing with admiration. "Would ye 'ave thought that yon lazy un could 'ave done that? Yon's not a bird, yon's a acrobat."

As we came into the yard of his cottage he said, "If ye'll allus keep yer eyes oppen ye'll see where t' birds beat all aeryplanes. Watch t' wild duck and see how she can throw 'ersen up perpendicular. Ye may git a shot at one of 'em as they comes doon river, but afore ye can pull second trigger, t'other duck has shot straight upards, like lightnin'. And when ye see t' kestrel hoverin' like a statue i' mid-air, ye —"

It was at that moment that his wife looked out of the door and called, in a tone that could not be denied, "John, your dinner's gettin' cold." He invited me to follow him with an inclination of his head, but I tapped my bulging pocket — and John understood. I can have my dinner indoors any day.

After finishing our sandwiches Raq and I made our way to Jerry's cottage. Whilst he was busy sawing some logs I told him of our morning with John Fell.

"Aye, but have ye ever tried to run and sing at t' same time?" said he.

I wondered at first what he was driving at. Then the marvel of it dawned on me. The lark had breath enough to both fly and sing.

"A thrush," continued Jerry, "sits on a branch and tak's it easy durin' his performance. So does a nightingale and a blackbird, but yon lark must have a good pair o' bellows to climb an' sing as 'e does at the same time."

I looked at Jerry admiringly. I had heard the lark and the thrush sing thousands of times, and yet I had never noticed how different were their ways of singing, nor had I thought of the lung-power which the lark's performance demanded.

For a moment I thought the old poacher was going to explain, then he seemed to think of something, and without giving any explanation, we strode off together towards a distant wood.

At the foot of a pine-tree he stopped, and I saw the dog putting his nose down to a few bleached bones.

"I thowt they might still be here," said Jerry, stooping down and picking them up. "They're all that's left of a woodpigeon a sparrer-'awk killed a few months ago." He put the bones in his pocket, and without offering any explanation set off in another direction.

By the side of a hedge he again searched, and again I saw a skeleton. The white head, with its chisel-like teeth, told me that it was the remains of some young rabbit.

"I wondered whether we should find it," said he, picking up some of the bones. "I saw it some time since. Carrion crows, I reckon, finished him off, poor little beggar."

Then he picked out one of each lot and handed them to me.

"Weigh 'em," he said, and as I handled them it dawned on me that this was Jerry's way of answering my previous questions.

"It is as light as a feather compared with the rabbit's."

Jerry nodded, and at the same time broke the pigeon's bone in two. "All birds bones is t' same."

"Why," I said, "it's hollow too, and the rabbit's seem solid. It's a wonder, though, that under the strain of flying these frail bones do not collapse."

"She rides in air and she rides on air — an' there's no sich things as punctures."

His words recalled to me my boyhood experiences. In those days I was the proud possessor of a "bone-shaker," a family heirloom. I remember the bumping solidity of those thick rubber tyres. Then, as I contrasted it with the modern "Dunlops," I marvelled still more at Nature's craft which provided the birds with pneumatic bones.

"All 'er front bones," I heard Jerry saying, "'ave air sacs in 'em, so there's no need fer a lark to run short of wind. Ye see, when we run, we've only oor lungs to supply us wi' air. But that little bird has air in her holler bones. I reckon she draws on them hidden supplies when she wants to give an extra trill or two."

"I must tell John," I said, as we moved off.

"Aye, an' give him these bones with my compliments," added Jerry.

CHAPTER
SIX

How Animals Eat

On the still morning air I heard the jangle of harness and the deep tones of command in which lay a caress. Alan was ploughing with Bonny and Darling. Raq and I at once made for the field and watched him slicing the purple-brown earth with his shining blade. He soon caught sight of us and waved a cheery welcome.

As I stood leaning over the gate, John Fell came along. "That's a nice sight," he said, "them gulls there followin' the ploo look like white yachts."

"And look at the metallic sheen on the rooks. No one seeing them in the air would think that they were such beauties," I said.

"Them gulls 'll soon be makin' fer their nestin' sites," said he. "Ye can see that their heads are quite dark. They're white or speckled i' winter."

Alan was at the far end of the field swinging round his plough.

"We shall get a nearer view of the birds when he comes down this end," I said.

John smiled, and waited. Gradually the horses came nearer, but the birds which fluttered around the driver so that he could almost touch them, left him as they neared us, and settled down in the middle of the field.

"I thowt they didn't like the looks on us," said John. "They'll wait till he turns round and then pick him up agin."

"Owt fresh?" asked Alan, as he tied the reins to the handle of the plough and left the horses to stand for a moment.

"Only the mornin'," said John, rubbing the noses of the horses. "Come and feel this," he said to me, pointing to Darling's nose.

The old Clydesdale knew that he was the centre of interest, and as I moved towards him I noticed his ears were erect and that he sniffed in order to get my scent.

"Did ye ever feel owt softer than that?" asked Alan, as my fingers wandered over the flesh above the lips.

"It's like the finest velvet." I murmured, and Darling tossed his head as though to tell me the examination had lasted long enough.

"I reckon," said Alan, looking at John to confirm his statement, "I reckon that a horse's muzzle is as sensitive as a blind man's fingers.

John nodded, and I felt that it was a most apt description. Darling's lips as they met my hand had the touch of fingers that were reading Braille.

"They've got to be," said the keeper. "A horse hasn't any hands, but them lips tak' their place. I reckon they can sort oot the grass they eat as easy as we pick and choose oor food."

Alan went over to the gate and beckoned me to look at the grass field over which Raq and I had walked.

"See how patchy the grass is? Ye can see it plain now that it's just a-startin' to grow. Them patches o' short-cropped grass tell ye that a horse is finicky aboot what he eats. They've bin feedin' a bit in that field, and ye can see what they've fancied and what they've left. An' when ye put their feed into the manager, the first thing they do is to search through it wi' their lips — not because they're greedy, but because they're cautious. Their lips are feelin' whether there are any stones or snags in it."

I could see for myself the truth of what Alan was saying, for Bonny had turned toward the hedge and her mouth was nosing amongst the dead herbage. Her lips rippled over the tips of the grasses, touching, gripping, sorting, each touch sending up delicate messages to the brain arousing either desire or disdain.

"How would ye like a pair o' ears like these?" I heard John say as he pointed at Darling's. "Look at 'em turnin' back'ards and for'ards. Only those of a hare or a deer can equal 'em. That dog o' yourn relies on his nose. Darling can listen to Alan wi' one ear when 'e's plooin', and at the same time listen to what's goin' on i' front of him wi' t' other."

"An' if ye want to judge a horse's temper," Alan added, "look carefully how he uses 'em. If he puts

41

'em back when ye go near him, he's generally ready to nip ye wi' his teeth. An' if along wi' these signs he shows a lot o' white in his eyes — let somebody else 'ave him, fer there'll be a hospital job some fine mornin' "

In the distance I saw Joe coming up from the fields in which his ewes and lambs were feeding, so I waited till he joined us.

"Owt fresh?" he asked, after a hearty greeting. I laughed, for this was Alan's first question, and I answered him in the words of the gamekeeper, "Only the mornin'."

"Ye're a bit clarted up wi' muck," said Alan to his brother, pointing at the mud which bespattered leggings and trousers.

"Aye," said Joe, "there were a ewe on her back in the ditch bottom. It's lucky I found her, fer she couldn't a turned ower herself, and she took a bit o' shiftin'."

"That's a nice bit of plooin'," he added, appreciatively, looking at the long straight furrows.

"It might be worse," said Alan, pleased with the compliment.

Iron ploughs were being used, a single and a double, and somehow, as I watched the steaming horses, and Alan throwing his weight to one side to correct the "pull" of the share, I was rather glad that the noisy "tractor" lay unheeded near the stackyard.

"How do you keep your furrows so straight?" I asked.

"Well," said Alan, "it all depends how ye begin. Ye must get your first furrow dead straight, or ye'll give a twist to the whole field."

"He puts poles up and doon the field in the fust 'set oot,'" explained Joe. "If ye keep yer eye on 'em, the ploo'll obey ye. Eye and hand work together, I reckon, and o' course Bonny and Darlin' both know their jobs."

"I remember," he said, pausing for a moment, "my father askin' one of our men whether he were plooin' straight. 'Aye,' he answered, 'fairly straight.' 'That won't do fer me,' my father answered; 'There's only two ways o' plooin' — straight and not straight.'"

While Alan and Joe were working John and I went to look at the black Galloway cattle.

How contentedly they lay, chewing the cud, their eyes, with the large lashes fringing them, drowsily half-closed, their noses moist and shiny! As we came nearer they snuffled loudly, evidently having caught the scent of a strange dog.

"That habit of chewing the cud is curious, isn't it?" I said to Alan, as we returned to the field. "I wonder why a dog doesn't swallow his food, and then bring it up to chew all over again as the cow does?"

"He bolts it," said Alan.

"All the more need for it to return and be chewed up again into little pieces," I said.

"Aye, that's reet now," said Alan, scratching the back of his head. "I must have seen thousands o' beasts lying in t' field chewin' their cud like them Galloways, and yet I've never thought of it afore. Why they do it, and dogs and horses don't, beats me."

"Sheep and goats chew their cud too, and I've seen the giraffe doing it. What do you think is the reason, John?" I said.

"Well, ye see, it's like this. When a coo eats, it passes its food into its first stomick. Ye can call it its 'crop' if ye like."

"How many stomachs has it, then?" I asked incredulously.

John laughed. "Three or four, I reckon. The first 'un is only a restin' place fer t' grass it swallers, or t' cud. Now it 'as to eat a rare lot afore it's got enough to satisfy itsel' and if it had to stand and chew all it needed, it 'ud have to stand fer t' best part o' t' day. That wouldn't matter so much i' these days, fer there are no lions or tigers lyin' aboot in t' grass ready to spring on't, but —"

Alan broke in, "It's as plain as a pike-staff. What did I tell ye? Them coos, or their ancestors, had to take in as much as they could in as little time as they could, 'cos they were in danger. Then, when they were full up, they retired to some quiet and safe spot and chewed it all up."

"I suppose there's not much feeding value in a small quantity of grass."

Alan shook his head. "A dog's food — meat mostly — is concentrated, and he can get a lot o' nourishment in a small portion. But t' coos must eat a lot o' grass to keep 'em goin'."

We left Alan with the plough and wended our way down hill. I looked at the Galloways again as we passed until Raq nudged me with his muzzle to tell me I was wasting time.

"Raq, old man," I said, "I wish I could eat my food and put it on one side to finish, as those cows do. Your mistress would tell you that I bolt my food like you do. But what a boon, when the gong sounds and I'm busy, to be able to rush to the dining-room and bolt the meal before the rest of the family arrives and then re-chew it later on! Besides; it would mean that I tasted it twice!"

Walking through the fields I thought of the hundreds that have to work in offices and in heated noisy mills. I smelt the freshly-turned earth, heard the joyous care-free song of the wren in the hedge, felt the cold buffet of the east wind tinged with the flavour of the hill-tops, saw again the pulling horses with gulls above. How I wished they could share my joys!

CHAPTER
SEVEN

Walkers, Hoppers, Dancers

"I saw ye ovver a mile away," said Ned, as Raq and I found him waiting for us, leaning over a gate.

"How did you recognize me at such a distance?" I asked. "Did you catch sight of the dog?"

The old man shook his head.

"I know yer walk," he said. "Everybody's got a gait o' their own. Ye can mask yer face and wear different clothes, but ye can't alter the swing o' yer legs or the way ye move aboot."

"That's true," said I, as we walked on, and I began thinking of the various walks of my friends. "John Fell," I said, "walks with slightly bent knees."

"He's spent 'is life i' the woods," said Ned, "and is allus bendin' under trees and bushes. Jerry, now, swings his legs from the hips, free and easy like."

"And what's my distinguishing mark?" I asked, with a smile.

"Ye've a tendency to swing yer toes in a bit," said Ned, without hesitation. "It's not exactly elegant, but it's the more nateral way o' walkin'. Stalkin'

46

aboot with feet turned oot like the 'ands o' a clock at ten minutes to two, strains the ankles and tires the feet. Fer easy walkin', either walk straight, or turn yer toes in a bit. I've done thousands o' miles in my time, and I knows what I'm talkin' aboot. 'Ave ye ever thowt 'ow different-like birds walk?"

"No," I said, "I don't think that I ever have."

"It's a fascinatin' study," said the old philosopher. "The art o' movin' aboot well repays a bit o' study."

As we reached the farm he said, "Whilst I go in and deliver these letters, just look around ye a bit, and ye'll see what I say is true enough."

Whilst Ned was away I took his hint and looked at the stragglers in the farmyard with renewed interest. The cockerel strutted about and seemed to be cultivating the "goose step," whilst no one could have mistaken the waddle of the ducks. Up in the fields the sheep appeared to walk on wooden legs. What a contrast between their progress and the easy swing of the dog that rounded them up!

Ned returned and found me watching a blackbird which was turning over the moist leaves under the shadow of the hedge.

"Well?" he asked.

"It is as you said," I answered. "Each has a peculiar walk all its own. I was watching that blackbird. It's a hopper. And there's a little black-and-white bird over there by the pond that walks like a human being."

"Ye're right," said Ned. "That pied bird is a wagtail, and as dainty a little critter as we have. She's the smallest bird, too, that walks. And, what's more, by the way they hop or walk, ye can tell summat aboot how they live."

As we walked along together the old man began to open out his theme.

"Whenever ye see a bird that hops, such as —"

Here he paused and looked at me.

"The sparrow, wren, or tom-tit," I said.

He nodded and continued —

"Ye'll know that it spends most of its time i' the trees and bushes. Walkin' ain't no good on thin branches. A hopper, ye see, is a bird o' the trees, and a walker is one that spends most o' its time on t' land."

As we passed through some stubble a number of rooks were diligently exploring some arable land. Most of them were in pairs.

"All walkers," I said to Ned, "with a bit of a sailor's roll."

"Wait a moment," he said; and as we watched, one of the walkers suddenly changed its gait, and hopping up to his partner, offered her some little dainty he had found. There was something of the comic about this sidling hop that made us both laugh.

"That's just a little bit o' the ritooal of his annooal courtship," said Ned. "It's hardly a hop; it's a'most a bit o' a dance. Did ye notice how he swung round to her?"

48

"I suppose walking was too staid a way of offering it," I said.

Ned nodded.

"I reckon it was. Some little bit o' joyousness at the thought o' the good time comin' put a — put a —"

The old man looked at me as he always did to suggest a suitable word.

"A lilt," I said.

"That's it," said Ned, enthusiastically, "I was seekin' fer a word that 'ad some chucklin' sway in it. It put a lilt into 'is black legs, and altered his gait from a stalk to a trip."

In the next field some plovers were doing marvellous aerial evolutions. One finally alighted and with meticulous care folded its wings.

"Another walker," I said.

"Aye, it both walks and dances."

"Dances?" I queried.

"Aye, dances for 'is dinner. Ye've seen 'em all standing, facin' t' wind?"

"Like tightly reefed yachts they look," I interrupted.

"That's good. Well, each is listenin' on its own particler bit o' ground."

"Listening?"

"Aye, listenin' for worms an' beetles movin' under t' ground. If 'e 'ears owt, 'e turns 'is 'ead first one side, then t' other, tryin' to lo — lo —"

"Locate?"

"Aye, that's it. Tryin' to locate the spot. When 'e's sure, 'e dances on t' ground until it comes oot. Down dives 'is bill and oot it comes."

"But why does he dance?" I asked.

"Worms is sensitive critters. An' a vibration yards off frightens 'im, so 'e comes up. Mebbe, 'e thinks a mole is after 'im — mebbe, 'e comes up in fright."

After we had said good-bye to Ned, Raq and I sat down on a knoll to let something of the sweet countryside soak in.

Away in the distance I saw the steaming horses in front of the plough. How steadily they moved along the furrows. They neither hopped nor danced, but how cleverly they planted their feet so as not to undo the work of the plough.

How leisurely and restful it all was. How slowly the boy moved behind the cows as they made for the byre. With what slow measured strokes the ditcher worked. No haste marked the thatcher as he put his rye straw on the cottage roof.

The sun began to sink and the cottage-windows glowed with mellow light. Along the lane I could hear the footfalls of the horses returning homewards — "Clop, clop — clop, clop — cloppity, click, clop."

The hoot of a car in the distance typifying all the rush and bustle of the town, brought me to life.

"Come on, Raq. We'll have to go. Back to the tyranny of the town again."

CHAPTER
EIGHT

Early Nests

"Have you found any new nests yet?" I asked Jerry, as we turned our steps towards the river. The land was still in the grip of winter, and the keen east wind had piled up the snow into deep drifts. Here and there could be seen holes into which terrified rabbits had plunged, and the signs of their search for food lay everywhere.

"'Ave a guess," said Jerry in answer to my question.

"A raven's up on the fell?" I hazarded.

"Aye," said my friend, "t' raven is nestin' up amongst t' crags yonder, but there's one I come across not far from 'ere. Like to 'ave a look at it?"

Jerry saw the answer in my eyes, and led the way to a corner of the field. Here the pine trees acted as a screen, and in the middle of a hawthorn bush whose branches were whitened with snow I saw a little cup of moss, and in it three eggs as blue as Italian skies.

"Hedge sparrow," I said delightedly, experiencing that thrill which never fails to come when one sees the first wild bird's egg of the season. "I wonder where the bird is."

Raq was nosing about in the bottom of the hedge, and he might have known what we were talking about, for as he brushed through he frightened the little bird out of the hedge. She settled on a branch for a moment, eyeing us a little apprehensively, no doubt wondering whether we had discovered her treasures. Then she uttered a sweet quick song and vanished.

"Was that a song of fear, joy, or entreaty?" I asked.

"I reckon she were a bit upset so had to oppen her safety-valve some'ow," said Jerry. "When we git excited I reckon we laugh or cry or —"

"Swear?" said I, looking at him severely.

"You say 'Hang it,' just as 'ard as I say 'D— it!' What the diff'rence is atween 'em beats me," answered the old poacher, with a merry twinkle in his eyes.

Again we caught sight of the little mouse-coloured bird. She was on the ground at the edge of the pine-wood, diligently turning over the fallen leaves.

"Ye can see by 'er bill she's no true sparrer," said Jerry. "Look 'ow fine pointed it is — it's not made fer dealin' wi' seeds, like the ordinary sparrer. It's made fer pickin' up small insecks. Watch 'er feed, and when she's done, ye'll 'ear 'er 'return thanks.'"

Every inch of the ground was being searched by the bird. Then, when she had finished, she raised her head, flitted to a neighbouring branch, and poured out her thanksgiving.

"Did ye notice 'ow 'she,' or mebbe 'he,' flipped his wings to his sides afore he started to sing? That's a little courtin' trick that 'e only does when 'e's lookin' after 'is bride i' springtime."

"I get terribly mixed up with all these sparrows, Jerry. Is this one related to the common town sparrow?"

"No, ye're thinkin' o' the little tree-sparrer that lays grey-streaked eggs in a tree. This un ye're lookin' at by rights ought to be given 'is proper name — 'hedge-accentor.' But t' common 'ouse sparrer I reckon is the queerest chap o' t' lot. 'E and 'is family actuoally go fer a summer holiday each year, like you and yer missus does."

"They migrate, I suppose you mean."

"No, 'e don't swank it like some o' 'em birds do, an' go to t' South o' France or Africa. He just leaves the town and follers you and yer missus into the fields. Ye can allus tell 'im from 'is country cousin — 'e's so sooty."

"Where does he feed?"

"In cornfields with his relations."

"An' if ye listen ye'll 'ear 'im tellin' them all kinds o' tall yarns o' the gay life 'e leads — until one on 'em asks why 'e's so dirty and bedraggled."

I laughed heartily at Jerry's description.

"Then when t' harvest time is over, 'e'll go back to town to 'is 'ouse top, an' swank agin to t'other birds o' all 'is doings. 'Such clean rooms, such

well-cooked food and the attendance excellent,'"
mimicked Jerry.

On our way back I asked Jerry if he knew of any
other nests.

"It is on the early side, but I think I know where
there ought to be one. It's not very far from home."

Together we made a bee-line for Jerry's cottage,
and in the field behind it, where the trees and
hedges meet, he began to search.

"What are you looking for?" I asked.

"A stormcock's," he answered. "There's allus one
just aboot here i' the fork o' a tree — not very high
up neither," he added.

At that moment we heard the shriek of an
enraged bird.

"That's him," said Jerry; "he's got the temper o' a
tiger. Keep still and ye ought to see summat."

I called the dog to heel, and in a moment we saw
a magpie leave the wood rather more hastily than
was his wont. His feathers, too, were not in that
apple-pie order which we usually associate with that
pied dandy. Behind him and above him followed an
indignant and wrathful stormcock.

All the vials of his wrath were let loose. No hawk
ever pursued a finch more relentlessly. Once, he
literally pounced down on his foe, and it was with
the greatest of difficulty that the magpie regained
his balance.

Jerry laughed. "That'll teach that old poacher not
to come nosin' roond a part o' the wood where a

mistlethrush wants to mak' a nest. I reckon we shan't have much difficulty i' findin' it."

A whistle proclaimed the fact that he had spotted it. It lay in the fork of a tree, about six feet from the ground.

"That's it," said Jerry. "Ye see that bit o' sheep's wool that is flutterin' from its edge? Ye nearly allus find that hangin' loose from a stormcock's nest."

Jerry gave me a leg up, and I had a look into it. There was just one egg, more resembling a blackbird's than an ordinary thrush's egg. The blue background and the brown mottling were clearer and more pronounced than in theirs.

"I should have thought that it would have chosen a less conspicuous place for its nest," I said, as I regained the ground.

"I allus think," said Jerry, "that bit o' loose wool is put in to give it a straggly, old appearance."

I looked at the nest again. Certainly the bit of wool, moving slightly in the light breeze, did detract from its appearance of newness.

We heard the raucous angry voice of its owner again.

"That bird don't need to hide very much, I reckon," said Jerry. "There's nowt 'll attack it wi'oot receivin' a warm welcome. He fears nowt. I've seen him drop doon on a sparrer-hawk that's bin restin' on a branch near by. I don't suppose the hawk knew that the nest was near. But he were soon enlightened, fer the stormcock jumped on him from above and knocked him clean off his perch."

As we reached Jerry's cottage I said, "If any more wild, cold weather comes, those eggs will stand a poor chance of hatching, don't you think?"

"Aye," said Jerry, turning to go in, "wi' all their bravery the mistlethrush must have a lot o' losses that way. Early nesters tak' great risks. But if they fail once, they 'ave another try a bit later. I've known a bird mak' three nests and lay three times afore she's managed to bring up a family."

"But what about that raven's nest, Jerry?"

"It's a goodish climb up t' fells an' I didn't know if ye'd think it worth it."

"Worth it?" I said, and seeing the eagerness on my face he put a "bite o' summat" in his pocket, and we set off again.

We seemed to have walked miles before we at last reached the fells. We stood in silence for a time, feeling the quiet grandeur of those rounded hills. Grey crags peeped out from last season's dead bracken, russet and warm to the eyes. Here and there, browsing on the scanty turf, the sure-footed Herdwick sheep stopped feeding to gaze at us with eyes that ever express wonder and surprise.

As Raq approached, they faced and stamped their forefeet at him, and then scampered to safer quarters. In the distance we could hear a shepherd giving commands to his dog by a series of staccato whistles. Only his whistle and the bleating of the sheep broke the eternal silence.

We walked along the base of the hills until we came to a great gully. On either side the rocks rose

56

sheer from the smiling stream that threaded its way through the glen. Some mighty giant had driven a wedge into those solid hills, and left it as a sanctuary for the buzzards, the ravens, and the lordly peregrine.

"Look," said Jerry to me, pointing upwards.

Far above us, with great wings outstretched, I saw a bird floating without effort in ever-widening circles. From the blue there filtered down a tiny "Mew." The wings never seemed to move, yet higher and higher that bird climbed towards the sun, till it looked almost like a speck of dust.

"Buzzard," said Jerry. "I knows where he'll nest, I reckon."

Once again, save for the rippling beck, the silence claimed the glen.

Jerry whispered to me, with something like awe in his tone: "It mak's me want to tak' me boots off and tread softly fer fear o' disturbin' the sleepin' peace."

Then we started to climb. Jerry was like a cat, and took it for granted that I was sure-footed also. Now and then the sound of our heavy breathing was broken by a dislodged stone that clattered its way down the crags and landed with a soft thud on the grass below.

"Never leave hold of a good grip until ye feel another just as sure," called Jerry from above, "and keep lookin' up. Never look down, or ye'll get a dizziness in yer head an' a swimmy emptiness under yet westcoot. We're nearly there now," he added cheerily.

Finally, I saw him seated on a ledge, and as I came below him, he caught my wrist in a strong grip and drew me up alongside.

"What a view!" he said, gazing down at the valley.

"Where's the raven's nest?" I asked abruptly.

"We can't reach it," said he; "but ye can see it easily enough." And looking in the direction in which he pointed, I saw on a neighbouring ledge the object of our climb.

It was rather a big erection of twigs, and lined inside with warm wool.

"Only two eggs," I said; "bluish in colour with grey-brown markings."

"She's not finished her full clutch yet," Jerry said. "Rather small fer the size o' the bird, aren't they?"

For a time we sat on our perch and let our legs dangle into space. Up from distant fields there floated the sounds of earth. We could hear the cackle of geese, the lowing of kine, somewhere a dog was barking, and we could catch the distant laughter of children as they played in a farm-yard. But every sound was glorified and beautified. It seemed as though each note passed through a filter of air and reached us with all raucousness and edge strained out of it.

Then, as we were about to descend — which to my mind is far worse than the ascent — we saw the owners of the nest high up in the air.

Sailing along together, unafraid of peregrine or any other foe, they came towards their sanctuary. We crouched flat on the rock.

Suddenly, in mid-air, one of the birds turned clean over on to its back. I could see its two black legs pointing towards the sky. For a second it seemed suspended in the invisible ether, then as it fell towards the earth it regained an even keel, and emitted a harsh "Squak-squak."

"Come on," said Jerry, "let's leave them to their antics."

As we rose the birds saw us and flung themselves higher in the air. We could hear their metallic protests being echoed in the glen.

"What made him do a side-slip on to his back like that?" I asked, as we found ourselves once more in the valley.

"It's a habit o' the ravens," answered Jerry. "I can't say no more'n that. Frosts may come, and the gales o' March may blow, but spring's i' the air. The hares know it, so do the ravens, and I like to think that these funny antics o' theirs is gratitude showin' itself i' spurts o' revellin' life."

CHAPTER
NINE

Parental Care

May is a busy time for all gamekeepers. It is the month when they collect the eggs from the pheasants' nests and put them under broody farm-yard hens, as the pheasant hen is not the best of mothers. This enables them to keep an eye on the young chicks, and guard them from a score of perils.

Knowing where John Fell kept his pens, Raq and I paid him a visit. I told the dog that he would have to be on his best behaviour or he would be sent home.

We found the keeper with a broody hen in a sack. He was kneeling down preparing the nest in which she was to bring forth her game brood. Half-a-dozen pot eggs lay in the nest. These I knew were to be given her at first to see whether she would settle.

"Will she sit tight?" I asked, as he prepared to place her in the box.

For answer John put her into her temporary prison, and the old hen, with careful steps, immediately covered the eggs, emitting as she did so, a few satisfied clucks.

"She'll be all right, I reckon," said the keeper as we left her. "I'll give her the real eggs to-morrow evenin' if all goes well."

"Wouldn't an incubator be easier?" I asked.

"Well," said he, as we entered his hut, "an incubator is all right, but to my mind nothin' beats a livin' bird. Ye see ye've the chicks to think on."

"You don't think very much, then, of mechanical foster-mothers?" I inquired.

John shook his head, at the same time drawing out an empty box, on which we both sat.

"Mechanical foster-mothers," said he, filling his pipe, "'as no eyes for emergencies. They can't give warnin' when the hawk's aboot, and they 'aven't any comfort to give to the ailin' chick. Most on us speak lightly of the owd hen, but when ye come to weigh up what she does — well, she's little short o' a wonder."

"She's a silly old thing," I said.

"Mebbe, she is, but if fer 'undreds o' years you'd a bin turned into a egg-layin' machine — expected to lay four or five eggs a week instead o' just layin' a clutch i' the spring same as other birds — you'd grow a bit soft, I reckon. Ye must allus remember she ain't livin' a normal life, and I say, 'She's a wonder.'"

Away in the woods we heard the harsh challenge of the cock pheasant. John jerked his head in the direction of the sound.

"To start wi'," he continued, "she gits no 'elp from him as a mate. She sits i' a small dark box fer three week, with no cock-bird to bring her dainty morsels to eat to while away the hours, and no bird wi' a golden voice to sing oot his love to 'er. But she just carries on wi'oot a murmur."

"That's true," I said, encouragingly.

"And when t' chicks arrive and she's simply bustin' wi' energy to tak' 'em oot and scrat up worms and seeds fer 'em, instead o' that, she's put in a coop, and there she stops fer another eight or ten week, wi' 'er head and neck fer ever stretchin' through t' front bars ready to give t' chicks warnin' if she catches sight on t' hawk or owt else."

"Don't you think she gets any pleasure out of it?"

"Her best time," said John, "is when chicks feel cold and run under her breast feathers and creep up t' side of 'er wings — and at night, when she feels the whole fluffy bunch underneath her, and the little beggars are cheepin' oot content and satisfaction. Then ye can see a prood look come inter her eyes — the look o' real mother'ood."

"There's a bird in yon woods that's interestin'," said John, pointing to a spinney that lay towards the river, "a woodcock sittin' on four eggs — least-ways she were on the nest a day or two sin'."

"I should like to see her," I said eagerly.

"Well," said John, "we can 'ave a try if ye like. O' course, she may have hatched oot — but ye must leave Raq behind. I shouldn't like to scare her."

A few minutes' walking brought us to the edge of the wood. John walked into it very carefully, though I noticed that he was not so silent a stalker as Jerry.

Crouching behind a mass of bramble, he told me to peer round cautiously. I did so, but for some time I could see nothing. Then from amidst a litter of beech leaves I caught sight of a pair of brilliant black eyes. Then the form of the bird took shape. I could see its long bill, something like a snipe's, but the plumage — a warm dark chestnut, with its stripes and bars exactly like the litter around it — held me fascinated.

I have marvelled at the beauty of the mallard-drake, and watched the shimmering loveliness of the kingfisher, but I have never seen more glorious tones than in that sitting bird. The richness of the browns reminded me of the silken pile on pan-velvet.

The woodcock did not stay after she saw that I had spotted her. As silent as a thought she was there one moment and gone the next. Then I saw four eggs, with a yellowish-creamy background and lightly mottled in ruddy buff, lying in a small depression on the ground.

"That was worth coming a long way to see," I said to John.

"And what kind of a mother does she make?" I asked, as we threaded our way out of the wood.

"One o' the best, I reckon," answered my companion. "As is usual wi' birds which build on t' ground, the little 'uns come oot o' t' shell already dressed, not bare like young sparrers and linnets. They 'ave their eyes oppen an' all, and can run aboot soon after hatchin'."

As he was speaking a very young rabbit rushed from the longer grass of the field, and, after hesitating for a moment at the entrance of a burrow, disappeared down it.

"There," said John, "ye have a young un that comes into t' world just like the sparrer and the linnet."

"Blind and bare?" I asked.

"Blind and bare," John echoed, "'cos he's born i' safe surroundin's, as you might say."

He said no more for a few minutes, but tramped steadily on in the direction of the High Barn on Alan and Joe's land. I refrained from asking any questions, for I could tell by the set way in which he was walking that he preferred to show me rather than tell me something.

Entering another small wood, he made for a clearing in the centre of it, and amidst the growing bluebells searched for something.

At last he stopped, and pointing to a small hollow in the green leaves, he said, "There were a young leveret sittin' snug i' there the day afore yesterday, but it's cleared oot. Howsomever, ye can see its 'form' where it hid itsen. I wish ye could 'ave seen it, for though a young rabbit and a young hare look alike, yet they are very different when born. We'll git

64

oot o' here, fer there are some sittin' birds aboot, and the less disturbin' on 'em the better."

Once in the open, John continued — "That leveret is born wi' his eyes oppen like the young woodcock, and covered wi' fur — the rabbit is born premature-like, naked and blind. D'ye see what I'm gettin' at?" he asked.

For a moment I did not reply. I was busy trying to link up the young woodcocks with the rabbits and hares. As I hesitated, John continued —

"Ye see, them as has nests i' the bushes or under ground aren't i' so much peril as the woodcock nestin' flat on t' ground — so the young uns are born blind and naked, and can tak' time to develop. The rabbit, ye see, has a nest that's hidden underground, and so its youngsters are born helpless."

Then I began to see the drift of his argument, "Oh! I see. A hare has its youngsters *on* the ground, and so must bring them into the world with their eyes open, their legs strong, and a suit of clothes, so that they can get away from their enemies."

John smiled and said, "Ye're larnin' all right, and now ye see that Natur' never does anythin' by chance, but that there's a reason fer everythin'. Now I'll answer ye whether the woodcock mak's a good parent."

"How far do ye reckon that woodcock's nest was from the river an' swampy land?" he asked.

"A good half-mile or so," I answered.

"Young woodcocks need marshy land fer their nateral food," said John, and looked hard at me.

"Well," I said, "the wood stands high and is as dry as a bone. They won't find much there."

"That's what I'm comin' to," said he. "That owd broody hen wi' her pheasant chicks that ye saw, ud scrat up food fer 'em if she were let loose. Sparrers and finches find insecks and bring it to the helpless hungry youngsters i' the nest. But the woodcock is a wonderful mother, for as dusk comes on, she actooally carries her young uns doon to t' marshy land. She doesn't carry the food to 'em, but carries 'em to the food." John paused, and I could see that he was gratified at the wonder which showed in my face. "Have ye ever seen a bird carry its young?"

I nodded. "Yes," I replied, "I've seen a swan with its tiny cygnets on its back."

"That's true," replied the keeper, rather grudgingly I thought, "but they get up there fer warmth and fer a lift agin a strong current, not to find food, ye know."

"And I suppose the young woodcocks climb on to their parents' backs and are carried in that way to their dining-room, aren't they?"

John smiled.

"No," said he, "the owd birds actooally grip the youngsters up wi' their legs. Some say as 'ow they carry 'em atween their feet. But I rather fancy that they carry 'em atween their thighs. I've heerd, too, that the owd bird puts her bill underneath 'em and 'olds 'em up — mak's a kind o' seat fer the little beggars to rest on, but I've never seen that part o' the carry-on mysen."

"And when once the young woodcocks are on marshy land, I suppose they stay there for good, do they?" I asked.

"Not at fust," replied the keeper. "As soon as dawn comes creepin' up, the owd birds bring 'em back to t' nest agin, or very near to it. But what are ye sighin' fer? Are ye tired?"

"No," I replied. "I was just thinking how tiring it must be, and how unselfish the parents are."

"Aye," said John. "There's not much we can teach 'em in that line."

CHAPTER
TEN

The Toilers

What a delight to potter about the hedges, knowing that at any moment some delightful little master-piece of craftsmanship may be discovered. In other words, what a delight bird-nesting is!

Passing a cottage I noticed a woman tipping up a box and knocking its contents out on the lane. It had been the brooding-box of a hen, for I saw quite a quantity of soft feathers.

No sooner had she left the unsightly pile than half-a-dozen birds pounced down on it, and a fierce struggle for the feathers began. A wren with scolding voice tried to brow-beat a tom-tit, and whilst they were haggling and quarrelling over a particular soft brown feather, a swallow deftly skimmed by and picked it up from under their very noses, carrying it in triumph into a cowshed.

Once a gust of wind carried one up into the air, and this gave a sparrow its chance to dance about in mid-air until he managed to secure the prize. I never saw such competition for feathers as there was that morning. When, returning later in the day, I examined the little pile of old nesting material, there was not a single feather left.

I lingered on the fringe of a wood where I thought it was likely that Ned, the old postman, might pass. Raq began digging away at a hole a few yards away. I noticed that at the entrance the earth had been scraped away, and lying about were numerous bunches of soft blue down. It did not take long to reconstruct the tragedy. A fox had found a nest of young rabbits and had evidently made a meal of them. The blue down was the fur pulled from the breast of the doe, in which the unfortunate youngsters had nestled. They, too, had known the joy of a feather bed! Wondering what made Raq continue to scratch at what I thought was an empty hole, I finally pulled him away, and, after searching, found at the far end one dead youngster. It had either died of exposure or exhaustion. It was blind, naked, and cold, and looked a forlorn, pitiable little object. Reynard had probably been disturbed at his feast, otherwise he would not have left such a dainty morsel behind.

In one of the trees, and built at the extreme limit of one of its drooping branches, Ned found me looking into one of the most exquisite nests that I have ever seen. It was only just finished, and stood ready for the mother bird's eggs.

"What nest is this, Ned?" I asked, and, pointing to a mass of soft iridescent silk, which made up part of the lining, I added, "And what's that stuff?"

"Thistledoon, I reckon," said the old postman, "an' ye're lookin' at a goldfinch's nest."

In the midst of the down was a lovely feather, glowing with bronze and green, which I recognized as from the neck of a cock pheasant. It gleamed like an opal in a bed of pearls.

"I wonder if the bird picked that up by chance, or had it a taste for colour?"

Ned did not answer me, but I knew that he too was silently worshipping at the shrine of perfect workmanship.

"Look at yon hairs, all woven neatly roond and roond, wi' not an end loose," he said.

"Beautiful," I said. "But where did it find them? You never come across a bunch of horse-hair lying on the ground."

For answer Ned walked across the field towards a couple of horses grazing at its lower end. Then he took my arm and led me to the gate-post. On it were a few hairs, caught by the rough edges of the wood.

"That's where them owd 'orses come to scrat and rub theirsels. Them li'l birds 'll search every post in t' field. Just think o' the 'ard work it means fer 'em, and that's nowt compared wi' a long-tailed tit's nest. Why I've found ower two thousand feathers used fer t' linin' only."

"In one nest?"

"Aye, and then there's moss ter find on top o' that — it must be real 'ard work findin' 'em all."

I marvelled at their patience.

Calling to Raq, Ned and I turned homewards.

We were about to move on, when I noticed that several birds were excitedly mobbing Raq as he ranged over the marshy land at the bottom of the field. I could see them swooping down noisily at him.

"Look at those sandpipers trying to eject my dog," I said, whistling him to heel.

Ned glanced at the birds and said, "Ye nivver saw a sandpiper wi' red legs like yon."

Meanwhile, Raq trotted back to me rather reluctantly, and the birds soon settled down again.

"Watch 'em use their wings," went on the old postman; "there's no bird can lift 'em more gracefully."

I focussed my glasses on one of them, and heard it give a call of two notes, which reminded me of the curlew's. "Too-ee, too-ee," floated up to us. Then it slowly raised its wings, so that I could see the light feathers underneath. In size it was bigger than a thrush, but it had the long, pointed bill of the snipe.

"What is it then, Ned?" I asked.

"That's a redshank," said he, "an' 'e lifts 'is wings like them owd pictures o' angels that ye see. I reckon when Raq started wand'rin' doon them medders, one on 'em saw 'im, and called up all 'is neighbours to mak' a massed attack on 'im. They don't like anythink prowlin' near their nestin' quarters."

"Have you ever found a nest of theirs?" I asked, as we sauntered down a lane.

Ned nodded. "They've got a real cunnin' way o' makin' t' nest in a tuft o' grass on t' ground. 'Ave ye ivver seen a woman usin' them curlin' tongs? Well, t' redshank gits 'old o' the long grasses wi' 'er bill, and as she sits on t' nest, curls 'em ower 'er so as to mak' a shelterin' roof. Ye can stand right by sic a tuft and niver know as 'ow the bird were sittin' there."

"Too-ee, too-ee," we heard the bird again.

"That means 'Good riddance to ye,'" said Ned to Raq with a smile.

"Did ye notice, when we was lookin' ower yon gate, what a lot o' work's goin' on in t' fields?"

"You mean the lad with the horses and plough or harrow, whatever it was?" I answered.

Ned shook his head, and pointed through a gap in the hedge. "Look down yon 'edgeside and count the number o' birds ye see. Not one on 'em is idle."

I did as he bade me and was surprised at the number of thrushes, blackbirds, larks, pipits, and finches flying busily about, whilst at the bottom of the field the rooks stalked about probing in the soft ground.

"Ye 'ears folk talk o' the beauty o' Nature — but there's more'n beauty to be seen. There's real 'ard work. Ye 'ears folks talk too o' the bounty o' Nature, but that's only part true. Nothin' drops inter the mouth o' birds or beasts, wi'out searchin'. They work 'ard fer all they git, I reckon."

The old man looked intently at me to see whether I was appreciating it all, and then pointed to a rook which was flying overhead. I could see the pouch under his tongue bulging with insects.

"They're allus at it," Ned continued. "In winter food is scarce, and they spend their days keepin' t' wolf from t' door. When food is plentiful i' spring, then they've housekeepin' duties. That there owd rook'll spend from March to Whitsun makin' tirin' journeys all the day long from t' rookery to t' fields. Whilst 'is mate sits on the eggs 'e's got to feed 'er as well as 'isself. Mind ye, she keeps 'im up to scratch. Then, when t' young 'uns 'atch oot, he's all them to feed as well, an' night-time is 'is only chance of a rest, pore feller. It's a full-time job from daybreak to sunset, not to mention the work o' nest-makin' and actin' sentry ower it, to save what's 'e's gathered bein' pinched by t'other birds."

I looked up at the old rook with a new respect, but Ned was drawing my attention to something else.

"Here's another busy feller," he said, pointing to endless mounds of soil. "'Ow would ye like to dig oot all them afore ye got yer dinner? That's t' mole's work, and when t' frost 'ardens t' ground, I reckon 'e's got to sharpen 'is spade an' work 'ard."

"But stoats and hunters in general seem to manage to get their food pretty easily," I said, thinking of the number of times I had heard the piteous squeals of rabbits.

"Ye only 'ear their successes; ye don't see the times they fail an' 'as ter go 'ungry."

He turned to take in his letters to a cottage which stood back from the road. "Sit doon, an' if ye've still any doubts, think aboot t' bees an' wasps and ants. If ye come across a lazy 'un, it'll be a dead 'un, or one that's on t' way there. In fact, they can teach you and me a lesson. The only bein' who gits some one else to feed and clothe 'im wi'oot doin' nowt, is man, tho' whether sic a feller ought to be called a man is oppen to question."

CHAPTER
ELEVEN

Nature's Adaptability

John Rubb and I were out for a day's fishing together. It is always an education to be out with him, for he has a store of natural lore and anecdote — only he must be in the mood to unfold.

Before us the river lay serene and smiling. For an hour or two I had whipped the water with little success. The river needed rain to scour out its bed and to revive the flagging spirits of the fish. True, I had captured a couple — but what are they when the surface is dimpled with rising fish, and the water so clear that they can detect your presence from afar?

Raq, too, enjoys my fishing. I wade into the water, and he seats himself on the bank, looking as grave as a judge. Every cast he watches with serious eyes. Then, as soon as I "strike," he is on all fours, showing a most lively interest.

If the line comes slackly out of the water, he settles down again to his quiet vigil. Should, however, the fish be hooked, and the reel scream out, then he rushes up and down, following the line's direction, and, as I finally take the fish from the net, evinces the liveliest satisfaction at the catch. I wonder whether he connects the landing of fish with the scraps of trout which he receives after they have been cooked, or does he enter into my joy?

From where I was standing I could see John's cast alighting on the water as light as a whisper. Then I saw the slight upward flick which signified that he had struck at a fish. Next moment, I saw that he was playing a beauty.

I went to see if I could be of any assistance in landing it, but I soon saw him handling a "pounder." He, too, came out of the water, and holding out the shadowed beauty, pointed to its golden brown, flecked with scarlet spots.

"They're bad catching to-day," he said, sitting on the bank. "You have to work hard for all you get."

I opened my creel and showed him my scanty total.

"They can see so well," said he. "Just have a look at that big one's eyes and you'll see that they are made very similar to ours."

"But can they see the same as we do?" I asked.

"Those who have studied the question," said John, "say that they are a trifle short-sighted. But the most interesting thing about their eyes is that they can keep the attention of one of 'em on a fly, and yet be watching you with the other,"

We chatted on, and I was quite surprised to see John reach for his bag. I had no idea that it was lunch-time.

As we munched each other's sandwiches John looked at a distant field, and so I did the same.

"Birds of the sea following the plough," he said. "By rights those gulls ought to be on the cliffs following the tide."

"But I've found their nests within a half-a-mile of the shore."

John nodded. "That's true," he said, "but all the same, the black-headed gull is fast becoming an inland bird. It's learning to find food, and finding it easier to get on land than by the coast. You'll find it visiting all the rubbish tips, and the worms turned up by the plough are easy prey. Have you ever seen them standing amongst a lot of plovers?"

"Often," I answered.

"Well, you watch carefully the next time," he said, "and you'll see a bit of fun. All the birds will be facing the wind. They don't like their feathers ruffling. Each little plover has his own bit of land, and here and there stands a gull. He waits as though he is not particularly interested in what is going on. The plover, however, is intently gazing at the plot in front of him and listening for the slightest sound of a worm under the surface."

"The gull tries to look bored and uninterested. But he's watching that bird like a cat watching a mouse. Suddenly the plover makes a move. Still the gull never winks an eye. Then the plover dives down his head, seizes the unsuspecting worm, gets it gripped in his bill, and —"

"Swallows it?" said I.

"Not a bit of it," John grinned. "That old thief, the gull, makes a dive too, and just as the worm seems about to disappear inside the plover, he diverts it down his own throat."

"And what does the plover do?" I asked. "Isn't there a fight or a protest?"

John shook his head. "Not as a rule. They suffer thieves as we do fools. But my point is, that the gull, being a sea-bird, hasn't yet learnt the art of finding worms for himself. If there's a plough in front of him — that's all right. It'll take him perhaps hundreds of years before he can cultivate as delicate a sense of hearing as the bird he robs — but like the trout, he's adapting himself to new conditions. Nature's wonderfully adaptable."

"He's another who's changed his habits," said John, pointing to a dipper who curtsied to us from a stone in midstream. "He was built for a land-bird, but decided he'd live in the stream. But we shall get no fish if we don't get on with the job," he said, rising.

By tea-time John had got a creel half-full of fish and I could only produce two small ones. As he lifted them out to show me, he said, "Have ye ever heard where these came from originally?"

"From these rivers, I suppose," I said, having just read about it; "and some of them did not find sufficient food so went out to sea and founded the migratory race of the salmon tribe."

John did not answer for a moment or two. Then he said — "Well, that 'ud take a lot of disprovin'

But that's not the latest idea, you know. It's thought now," he continued slowly, "that the ancestors o' these trout had their home round about the shores of Northern Africa and Southern Europe."

"How did they come to the English rivers, then?" I asked.

"Of course," said he, "we're speakin' of thousands and thousands of years ago. In those days, fish couldn't live in England for great glaciers covered the land. But as soon as these thawed out, then the Southern fish came northwards, and liking the flavour of the waters, and finding plenty of food in the rivers, they colonized the whole of 'em — and here they've stopped."

"That would be a great change in their lives and habits," I commented.

John nodded. "Nature takes its time to change, of course. Nothing happens sudden. But it's my belief that everything alive has the power to adapt itself to new surroundings if you give it time. Ye remember that time ye were on Ailsa Craig photographing the birds, you came across the guillemot, didn't you?"

"Thousands of them."

"Well that bird with its stream-line body, lengthened neck and webbed feet sprang from the same stock as the plover I was talking about. The plover has become a stunt flier, but the guillemot doesn't need wings much, and so has become a fine swimmer instead. Its whole body is shaped for cutting easily through the water."

"I wonder how many centuries it has taken to mould such a land-bird into a sea-bird?" I said.

"Thousands and thousands of them," John replied. "Nature takes her time but turns out a finished article at last."

And so we fished until the sun sank like a crimson disc behind the hills. I looked into my creel. There was not much to show for all my labour. It was almost empty; but the day had been a great success.

"Can't see what you can like in fishing!" is the refrain of many of my friends.

But I had seen the wheeling flight of the rooks, and watched the wood pigeons go to roost in the pines. The little water-vole had sliced a reed and boldly swum across the river. I had laughed at the rabbits as they played "Tig" in the sunshine, and seen the kestrel "quarter" the fields. The kingfisher had sped by like a jewel, and the heron had watched me from afar. Above all, I had heard the curlew. No longer was it merely plaintive or querulous in its note. Something else was creeping into it — something joyous, something overflowing.

As John said as we turned our faces homewards —

"That bird seems to have swallowed a stream and kept its gurgle in its little throat."

CHAPTER
TWELVE

The Weasel Moves in

As the sun poured down, I took refuge under the lee of an old stone wall dividing two fields. Raq sat in front of me giving me profound and sentimental looks.

Grateful and comforting was the shade of the wall, and the words of the shepherd came to me, "Hedges are a'reet and very fine to look at, but there's nowt beats a good wall. It's the best protection fer sheep from bad weather, an' if it falls doon, the stones are there reet on t' spot fer ye to build up agin."

I delight in these old walls, and have sat by the hour watching some old country artist build one. At one time I thought that it was the easiest of jobs to put one stone upon another, trusting to their weight to keep them standing firm and secure. But I soon learned that certain stones were looked upon as key-stones. They deftly locked the looser ones together. This was an art that had to be learnt, and many of the wall-builders had got the secret from their fathers. It had been passed on from one generation to another. To-day, however, "wallers" are an ever-dwindling race, and perchance in a few years' time the art of "walling" will almost be forgotten.

To walk along by such a wall in leisurely fashion has a joy all its own. How Nature drapes its bareness in green moss and grey lichen! Here the stonecrop has managed to find enough substance into which to cram its roots, there a small fern throws out tender fronds. In its crannies, robins, tomtits, wagtails, have found a shelter for their young — the weasel and stoat know it as a sure refuge.

And so in its shadow we rested, whilst the lark sprayed the corn-fields, and the pipit twittered sweet melody as it fluttered down from a tree.

I then made my way up to the field where the gamekeeper was rearing young pheasants. An under-keeper told me that I had just missed him, and directed me to where he thought I might find him.

"There's a lot o' rats in a wall 'e wants to clear oot, and I think 'e's takken some poison wi' 'im," said he.

I knew the wall he referred to. It was one that ran down from the uplands towards the farm. When I got there I saw him crouching behind a bush, and he motioned me to come quietly, and to keep Raq down.

"What are you waiting for?" I whispered. "I thought you were putting down poison."

"I reckon there ain't much need fer't." He pointed to a gap in the wall. "I were walkin' along 'ere quiet-like when I caught sight of a weasel comin' doon side o' t' wall. She were carryin' summat in 'er mouth, and at first I thowt it were a rat. But when she got to that 'ole, she waited, and I saw that it were one o' 'er young uns. She popped it in, and then went back same way as she'd come, so I'm a waitin' on her," he said grimly.

"But you're not going to shoot her, are you?" I asked.

John did not answer me, but simply kept his eyes fixed along the side of the wall.

About five minutes later, he pressed my arm. "'Ere she comes," he whispered.

For a moment or two all that I could see was moving grasses. Then, as she came into full view, I saw her carrying what looked like a small kitten in her mouth. She had hold of it by the back of its neck, and when she reached the opening in the wall, I saw her pause. Then she slipped in, carrying the youngster into its shadows, as John had described.

"There'll be no need o' me puttin' rat poison i' this wall now," said John walking away. As we followed him he gave a low chuckle of satisfaction.

"What is it, John?" I asked.

"Well, ye see, you weasel 'ad 'er nest some distance away. I calkilated she were away twenty minutes fetchin' each youngster. When I seen what she were carryin' I knew as 'ow she'd mebbe bin disturbed, or p'raps all this rain were soakin' in't, and she feared that it 'ud drown 'em where they were."

"So she'll make another home in that dry wall?" I asked.

John nodded, "An' that's where t' rats is," he said, and again he chuckled.

"How many youngsters will she have to carry?"

"Mebbe five or six, and as they git bigger and 'ungrier, there'll be some fine carryin's on i' that wall. As soon as them rats sniff their new neighbours I reckon a shiver o' fear'll run through t' lot on 'em. Some on 'em 'll clear oot reet away, but a lot on 'em 'll have young fam'lies to think on."

"But," I said, "even a weasel can't eat the lot."

John smiled. "It's not what she eats as matters so much. It's the fact that once she starts a killin' she goes on killin' till she's tired oot — then she starts ower agin. Ye can tak' it from me that in less than a month she'll 'ave cleared 'em oot, so I'm not wastin' me poison on 'em."

As we looked over a gate into a field which bordered the road, his keen eyes detected a couple of hares lying in the short grass. I only saw one of them, until the other slightly raised her ears, which were only visible for half-a-second. Then she lowered them again very slowly, as though she knew that sudden movement might attract attention.

Raq had been standing by my side looking longingly into the field. Probably the slight breeze had brought their scent to him. But I had restrained him from investigating.

"Send 'im in," said John, "'e'll do no damage, and I want them hares shifted from this field — it's too near t' road."

"Afraid of poachers?"

The gamekeeper nodded. "Ye never know who might be tempted when they're so 'andy — better to scare them to somewhere quieter, I reckon."

The dog needed no urging. As soon as he had run a score of yards, it was a fine sight to see both hares glide from their "forms" so quietly that they seemed to be part of the ground on which they were running. The one Raq was chasing soon sensed his slow speed and went off in almost leisurely fashion. After a few moments I whistled to the dog to return.

"I allus likes to see hares lyin' aboot like that," said John, as we waited for the dog, "fer it shows that nobbudy 'as bin rangin' ower t fields, especially when ye see 'em lyin' in t' middle on't, like yon two were."

"And what would you have thought if they'd been lying in a corner?"

"Nowt much," said John, "except mebbe that owd John Graham had kept passin' through to have a look at 'is sheep on t' fell yonder. 'E'd never meddle with t' hares, I'm sartin', but it 'ud be enough to mak' 'em keep oot o' 'is reglar track."

Raq returned panting and hot only to start chasing a rabbit.

"Do foxes do much damage to your hares, John?" I asked.

John shook his head decidedly. "Not much. T' hare is ower fast fer 'im, and 'er nose scents 'im oot even if 'er ears don't tell of 'is comin'. 'E's no match fer 'er, 'specially as she allus runs uphill, if 'e does chase 'er. She knows that them back legs of 'ers push 'er upwards so quick that he's winded. The fox that keeps on chasing 'er on steep rising ground 'as some 'ope. But foxes does a lot o' damage amongst t' leverets i' springtime."

We were crossing a stubble field as I spoke, and John stopped and pointed to a thistle.

"One o' the great dangers to them young hares is when t' farmer rolls the young corn. Its t' easiest thing in t' world for t' heavy roller to go ower 'em and t' crush t' life oot on 'em."

"I suppose they are so like the colour of the earth that you can't blame him for not seeing them? But what about that thistle?"

"A few months sin' I got word that t' farmer were goin' to roll this 'ere field, so, knowin' as 'ow a leveret or two might be lyin' in it, I thowt I'd 'ave a turn roond it afore t' damage were done, and I found two on 'em just by yon thistle. I took Nell, our setter, wi' me. She soon stood an' 'pointed' where they were, or mebbe I'd never 'ave seen 'em."

"And what did you do with them?"

"I put 'em a few yards ower t' hedge in t' other field."

"I reckon they yelped a bit when you were carrying them," I said.

John shook his head. "I didn't carry 'em — I trailed 'em to where I left 'em."

The keeper saw my astonishment and smiled.

"Ye 'member my tellin' ye what a grand nose an owd hare as. Now if I'd 'ave picked 'em up i' me hands as you said, I don't say as 'ow the doe wouldn't 'ave found 'em agin, but there would 'ave bin a big gap atween where I found 'em, and where I put 'em, an' —"

"Oh, I see," I said. "You took them by the ears and slithered them over the ground all the way, so that she'd follow their scent?"

"That's it noo. An' when t' owd hare came back to where she'd left 'em, she'd soon find the scent trail with that wonderful nose of hers."

Returning to the rearing field we passed some barbed wire. On one of the spikes I saw something hanging. On examination it proved to be a young bird that had hardly learnt to fly.

"Poor little beggar," I said, "it must have rushed full tilt at this wire and impaled itself. Look, John, the spike has gone clean through its neck."

The keeper walked over to it. "Nay, that li'l bird was put on that spike. It didn't crash on't."

"You mean that some one has deliberately skewered the wee mite?" I asked angrily.

"Some *thing*, not some one, 'as done it, I mean. Ever 'eard of a bird 'as is called a shrike?"

I nodded. "I've only seen one once. It's a bird about the size of a blackbird — mostly grey, black and white, isn't it?"

John nodded. "That's it. They're not common i' this part o' country. They're plenty down south. Ye can't mistake it, fer it 'as a broad black band o' feathers runnin' across the eye. 'E's the rascal as skewered that young bird."

"What does he stick it on the wire for?"

"Well, ye see, a hawk takes the bird to its killin'-mount, stands on it, holdin' it down wi' its talons, then tears it to pieces. A shrike is a bit too weak in t' legs fer that, so he finds a bit o' wire, hooks it on an' carves away. Some folks call it their larder, but it's not exackly a storin' place."

John left us, so Raq and I turned our steps homewards, pondering over all that he had told us.

CHAPTER
THIRTEEN

Fine Feathers

Whilst Ned went up to deliver his letters, Raq and I skirted the old orchard to see what was about. A woodpecker flew away at our approach, screaming out derision. Then I caught sight of a flash of pink. I thought at first it was a cock chaffinch, but a closer view revealed a bullfinch. He was a beautiful bird, such a contrast to his drab mate, coarser in build than many of the finches, but wonderfully active. His glowing breast appeared more fiery than the chaffinch's by reason of his black cap and other sombre colouring.

"I've been watching a cock bullfinch," I said to Ned when he returned. "Is he a favourite of yours?"

"Yes and no," he answered with a slight smile. "As a bird-lover, yes. But as a gardener, no!"

We moved cautiously, trying to get another glimpse of the bird.

"I've an easy round this mornin'," said Ned. "Let's 'ave a few minutes' sit on this tree-trunk. Mebbe we'll git another sight of 'im."

Raq came and squatted in front of us, wondering, no doubt, what we were planning.

"When they're movin' in and oot o' t' blossom," said the old postman, "they look like a red fire in the snow. 'Ave ye ever seen a pair of 'em sittin' on a branch touchin' their bills together? They kiss like a couple o' daft sweet'earts. Ye can't 'elp but be taken up wi' 'em."

"Now you are talking as a bird-lover, Ned."

The old man nodded. "But there's me fruit-trees to think on. Ye see them birds eat a lot o' insects, 'specially them as is curled up i' fruit buds. They not only goes into t' buds and pulls oot the grub that is lyin' inside, but pulls to pieces buds that 'ave nowt in, just to see if there is anythink in. That's when t' damage is done. One day yer tree is a pictur' o' delight. Next day ye may see scores and scores o' yer blossoms lyin' like shrivelled snowflakes on t' ground. That's when they mak's yer mad."

In spite of a long wait we did not see the bullfinch again, so we moved on.

"Well, which would you rather have, Ned — fruit or bullfinches?"

Ned was not slow in answering. "Oh, give me t' bullfinch every time. But fer them as wants their orchard to pay their rent, well, ye can 'ardly blame 'em if they tak' strong measures wi' 'em."

By the side of the lane stood a thick thorn bush which Ned pointed to.

"There were a bullfinch's nest i' t' middle o' that a year or two sin'."

"I've never been able to find one. I wish I had seen it."

"It's a bonny nest o' grass and small twigs, with a nice linin' o' hair, and sometimes a few feathers."

"And what are the little ones like when they come out of the nest?" I asked.

"Ye've seen a nest o' young robins, 'aven't ye? Well, both parents are t' same colour and are easy to see, ain't they?"

I nodded. "But the youngsters aren't a bit like their parents," I said. "You'd think some spotted brown bird had visited the nest and substituted her own eggs, wouldn't you?"

Ned smiled. "Ye're right. Howsomever, it's a great blessin' t' young 'uns are dowdy, fer they 'aven't much sense at first, and no experience of a 'ard world. If they'd fiery breasts like their parents, it 'ud call attention to 'em. When they've lived and learned a bit, then Nature begins to dress 'em up, an' —"

"And the young bullfinches?" I interrupted, thinking he had forgotten all about it.

"I'm comin' to that," he said a bit curtly, "I 'aven't forgotten. What I was goin' to say is that where ye git bright-coloured parents, there ye git drab-coloured youngsters as a rule, though there's exceptions o' course to everythin'." He hesitated here, as though he had not made up his mind how to continue.

Passing under a bare ash-tree a chaffinch rang out his spring call.

"Thank ye kindly," said Ned, looking up at the bird, "ye've just given me the idee I wanted. Now, where t' cock and hen are diff'rent, t' cock a dandy and 'is mate a sober-side, then ye'll find young uns at first all tak' after t' mother. Young chaffinches are pretty plain to look at, and young bullfinches too. Ned put plenty of emphasis on the last few words. "Youngsters are plain-lookin' like their mother — no red breasts — and Natur' even tones doon their black 'eads to brown. D'ye understand? I could 'ave put it better if ye 'adn't interrupted me when ye did."

"Sorry, Ned," I said apologetically. "I get so interested I forgot about butting in."

The old man squeezed my arm. "That's all right noo. I know I went a roond aboot road to git there. As we git older, I reckon we're in danger of gettin' crankier."

As Ned went with a parcel to a farm an old rooster strutted past.

"There's a big difference in the appearance of that cock and the hens," I said when he joined me.

Ned nodded.

"But not much between them two," pointing to a pair of rooks. "It 'ud puzzle ye to decide which is the cock and which is the hen."

We watched them for a moment or two, and then saw one bird sidle up to the other with odd dancing steps.

"That's the cock right enough. He's p'raps a shade bigger, but it's difficult to tell," said my friend.

All the way along Ned went on pointing out to me the varieties of plumage.

Here the jet-black blackbird, with orange bill, looked very smart against the more sombre dark brown of his mate, her bill merely the colour of horn.

There, a couple of thrushes flitted in the bushes, but Ned could not decide from their markings which was the master and which was the lady.

A robin sang on a hawthorn, and he asked me what colour the hen was.

"Brown, and its breast flecked with small stripes," I answered, purposely giving the answer he expected.

He shook his head and laughed.

"Nearly everybody thinks that," said he. "The youngsters are, but no one can tell the male bird from the female. Both on 'em have red breasts."

I felt, as Ned spoke, that he had given me a new interest, as I strolled about the countryside, to tell by its plumage whether the bird I saw was the cock or the hen.

"Are there many pairs of birds coloured alike?"

He thought for a moment, and then said —

"Oh, aye! A goodish few; though in many cases where the colourin' is the same, the hen bird's feathers are a bit toned-down like — they haven't the bloom that the cocks 'as. Swifts, whitethroats, nuthatches, bluetits, nightjars, sandpipers, flycatchers, house-martins, kingfishers, thrushes, wrens, barn-owls — oh! there's heaps on 'em. If ye'll go on a bit wi' yer investigations, ye'll find plenty o' food fer thought. There's first of all what I calls the law o' compensation."

He knocked the ashes out of his pipe and looked round on the fields with a meditative kind of air. I waited for the philosophy which I knew would follow.

"There's the thrush now, wi' a voice like an angel's pipe — but he ain't much to boast aboot in the way o' looks. But that old peacock up at the 'Hall' is like a walkin' rainbow; and yet 'is voice is like a 'undred tom-cats all caterwaulin' at once. Then, there's the woodpecker, nowt to listen to — just laughs wi' a daft kind o' chuckle; but he's a smart dandy in green and red. Nobody either 'ud look at a nightingale; but he's got all the colour in his voice. See what I mean by the law o' compensation? Everythin's got a redeemin' feature."

"And so have all of us, fortunately," I said.

"Wi' birds that have no song to charm their mates," he continued, "you'll often find that Natur' decks 'em out wi' sparklin' colours, but I rather fancy that the highly coloured uns have got to pay dearly for their decorations."

Ned looked at me to see whether I grasped his meaning; but as I looked puzzled, he said —

"Drab birds can sit on their nests in the trees and bushes. They can listen to their mates as the wind sings its lullaby through the branches, and tak' a sartain amount o' interest in what's goin' on aboot 'em. But the bright-coloured uns can't.

They've to tuck theirselves away in the darkness. The woodpecker goes into its hole in a tree; the kingfisher sits away up a tunnel i' the river-bank; the bright-coloured shelduck nestles up a rabbit's burrow. Fine feathers mean dark nests and a hermit's life."

As Ned finished there was a scuffle in the hedge, and the next moment Raq flushed a gorgeous cock pheasant. With a mighty whirring of wings the bird sailed into a neighbouring wood, whilst the spaniel watched his flight with blinking eyes.

"Fine feathers make poor husbands in his case," I said, pointing to where the bird had disappeared.

"That's truer than ye know," said he with relish. "So many o' them highly-decorated birds with drab mates want partners, but don't fancy the work of a family. They mak' husbands, but not good fathers."

"I know that's the case with the mallard," said I. "He's a dandy right enough, but soon deserts that dun-coloured wife and her pale-green eggs, but I don't know of any others."

"They're mostly game birds," said John. "There's the black gamecock and his grey hen, the capercaillie o' Scotland, and the grouse o' the moors. The hens do all the work o' sittin', and their lordships do all the struttin' an' showin' off."

"Fine feathers evidently make fine philanderers," said I.

Just before we parted Ned said to me —
"Ever heard of the red phalarope?"

"Never," said I.

"It's a bird that doesn't nest i' this country 'cept in the far North. I've never seen it mysel', but a naturalist feller were tellin' me on't."

"What is there interesting about it?" I asked. I knew he had some amusing idea tucked away in the back of his mind, for I saw a twinkle in his quiet grey eyes.

"Oh, well," said Ned, "in this case the HEN phalarope is the ornamented one, and her mate keeps hissel dressed soberly."

"That's more like human beings," I said, with a smile. "It's the fair sex that blazes forth in purple and fine linen and we —"

"Aye, that's true enough," said Ned, as he turned to leave me. Then he paused for a moment, and said —

"But don't forget it's the drab-coloured cock i' their case which does all t' housework. Sits on the eggs and tends the young uns, whilst his skittish partner decks hersel' oot and has a gay time."

He strode away in the direction of his cottage, and I called after him —

"And the moral?"

"You ask yer missus," he said, with a chuckle, as he vanished into the wood.

CHAPTER
FOURTEEN

The Corncrake

When Raq and I were sitting waiting for Ned I listened very carefully to the sounds we usually do not seem to have time to notice.

The grasses on top of the bank stirred in the slight breeze, rustling with intriguing music.

Then, when the air was still, I caught the undertone of a myriad unseen wings. Each little insect, as it hovered before red campion or stitchwort, droned out with its lace-like fans a certain note. Every fly had its own note. But my ears were not sensitive enough to distinguish the varied tones. All I could hear was the massing of those individual "hums" until they reached me as one soft purring drone.

Then, supplying a new theme, a great bumblebee tumbled blunderingly from one clover flower to another, swaying down the purple head until it almost bowed itself to the earth. Ever and anon the field-cricket took up the part of the fiddler and scraped his sheathed legs against hard wings.

In the grass, little chaps with long tails squealed with mouse-like voices, and once, from a neighbouring wood, there came the shriek of a rabbit in distress. As I sat and listened I fancied that Nature was playing her bagpipes. Beneath the varying notes of the theme there lay that continuous drowsy monotone of the insect world.

I wished my ears were delicate enough to distinguish each individual player's note. Then I remembered that I lived in a world which holds not only the music of the fields, but the hoot of the taxi and the crashing rumble of the wheels of commerce.

I was recalled from listening to the field's orchestra by hearing in the adjacent meadow — "Crake-crake."

Raq heard it too, but the only sign he gave was to be seen in the slightest of twitches with his right ear. Also his nose veered a point in the direction from which the sound proceeded.

"That's a corncrake, old man," I said excitedly, "Let's see if we can spot him. I wish Ned would come."

It was at this moment that Ned, the postman, appeared on the scene, so I told him our plan.

There was about half an acre of fairly long grass, and beyond this the field was cropped rather closely.

"He can't get away without my seeing him," said I to Ned; "he's bound to fly over the barer ground."

"Don't be too sure," said the old man; "they're pretty 'fly' are them 'crakes."

So, leaving the dog with Ned, I began to crawl towards the grass which screened the bird.

"Crake-crake," once more came the sound. I turned for a moment and winked at Ned as though to say, "He's still here, you see."

Soon I was on all fours stealing gently towards the corner from which the crake had called. I proceeded cautiously, keeping a wary eye on the open sides of the field, so that I might get a good view of a bird I had never seen at close quarters.

I had only a dozen more yards to crawl before reaching the end of the plot. Very carefully I advanced, at any moment expecting the crake to take to its wings and reveal itself.

But no bird showed itself. I turned towards Ned, and he immediately shook his head and called —

"It's still there."

Then, in mockery, there came from the other end of the field, "Crake-crake." The bird had actually slipped quietly past me, and I had not seen the grass move nor heard a rustle.

Determined to see it, I turned to retrace my crawl, but Ned called out —

"It's no good. Ye can keep on doin' it all day, and he'll fool ye at t' finish."

Reluctantly I had to acknowledge defeat. Had I loosed Raq he would very soon have flushed the bird, but I felt it was not quite playing the game, though he was just quivering to hear me say "Seek!"

I moved on with Ned, and, just as we were leaving the field, I turned to give a last glance at the hiding-place. From the farthermost end I saw a light brown neck, streaked with shadowed stripes and surmounted by a wedge-shaped head, craning up over the grass-tops. He was evidently enjoying his triumph.

"Have you seen one before?" I asked Ned, who chuckled as he saw the bird eyeing our departure.

"Aye," he said; "often enough. It's just made fer goin' through the grass quietly. Body shaped like a wedge. Long legs which it uses well, except when it flies. Then it dangles 'em like a waterhen. They look a'most as though they're broken."

"Some people say," I said, "that the crake has the art of a ventriloquist."

Ned snorted. "Some people 'll say owt," he said. "The curious thing is though, that its 'crake' allus seems to come from the very spot yer lookin' at."

Ned went along to deliver some letters, so I sat by the roadside and ate my lunch.

Raq kept running up a small path, peering at something and then returning with a look which said, "Come and see what I've found, master."

I made him wait until I had finished, and then went with him.

Up one of the smaller paths I saw him peering at something on the ground.

In a rising bank before us was a hole about the size of a bucket. In it was a quantity of what looked like grey cigar-ash, but also reminded me of thin layers of paper.

"Wasps' nest, or has been. But where are the wasps, old Nosey Parker?" I said to Raq.

"Some boys must have dug out a wasps' nest here, Ned," I called as he came round the corner.

Ned shook his head. "I saw it as I passed this mornin'. Lads never did it." He picked up a piece of bark which lay on the ground, and pointed to a track underneath it clearly printed in the clay soil. It was far too big to have been made by a fox. There were three very deep indentations made by claws, and two others much shallower which could only be seen after close scrutiny.

"The feller that made that track dug t' nest oot."

"Badger," I hazarded.

Ned chuckled. "Aye, I reckon it were them deep, bear-like claw marks as gave ye t' clue, eh?"

I poked about the débris and came across scores of dead wasps. Raq was once stung on the nose by one, so that when he saw their yellow bodies he sat down on his haunches looking as solemn as though he were at an inquest — and kept at a respectful distance from them. Every now and then he licked his nose as though the memory had not been forgotten.

"I didn't know that badgers were fond of wasps though of course I know they will dig out the nest of the wild honey bees."

"There's nowt 'e likes better than the young white grubs o' t' wasp — it's 'is fav'rite dish, an' 'e don't tak long in gettin' doon to 'em when once 'e's sniffed oot where they are. As a rule t' nest ain't far in t' bank, and wi' them strong spades o' 'is, 'e soon gets 'is meal."

As Ned spoke I could easily picture the scene. How often, in years gone by, with a gunpowder squib, have I doped and poisoned the thousands that lived in such a hole! How often, when I have been in need of the young wasp grub as fishing bait, have I tried to dig the nest out in the daytime without the aid of the squib! When once the nest was brought to light, it was policy to cut and run as quickly as my legs could carry me. Even then — But that is another story.

With these memories flashing through my mind, I said,

"But don't the wasps sometimes sting the badger?"

"'E doesn't seem to mind 'em," said Ned simply, and, picking up a few of the dead wasps, he asked, "What-d'ye think killed all them wasps lyin' there at t' bottom of t' hole?"

"The badger dashed them down with his paws as they tried to get at his nose," I answered promptly.

"But there are 'underds o' dead uns. 'E couldn't 'ave killed all o' them, and, mind ye, it 'ud be night time an' all, or p'raps dawn."

I certainly felt that the badger must have been exceedingly lucky in his downward smashes to have slain so many.

Ned saw that he had raised a doubt in my mind and smiled.

"I 'ave 'eerd it said as 'ow t' badger 'as a very strong smell — ye can call it a stink; that 'ud be nearer t' mark I reckon — and that t' smell poisons t' wasps, and if it don't kill 'em ootreet, at all events it paralyses 'em. I won't say 'as 'ow I can prove it, but I can tell ye that when I come across this nest the mornin' after it were done, there were 'underds of wasps crawlin' aboot on t' ground as though they'd 'ad too much strong drink.

"Well, that takes some swallowing, Ned," I said as we parted; "but I can't for the life of me think of any better explanation."

CHAPTER
FIFTEEN

The Cuckoo

To my delight, John sent me word that he had seen a cuckoo's egg in a hedgesparrow's nest.

So collecting all my camera baggage, and calling Raq, I set off with high hopes.

It was a beautiful morning. Everywhere the red and white may-blossom burdened the air with its heavy perfume — almost too languorous for May, and suggestive more of drowsy June.

In the woods the wild hyacinth draped the roots of the big beeches and elms in bluish purple, the wild anemones shook their bells with delicious shivers, and the primroses starred the dead bracken with pale beauty.

As I cut through a spinney, Raq "pointed" at the burrow of a rabbit. Evidently the tenant was not very far away, to judge by the hectic wagging of his tail.

Then from the large hawthorn-tree immediately above him, two brown birds launched themselves. I was so close to them that I caught a glimpse of lustrous eyes. So silently had they emerged that not even the dog had caught either the flutter of a leaf or a ripple in the air. They flew light as a thought, and with muffled wings vanished from my sight.

"Tawny owls," I said to Raq. "I must ask John about them."

As I neared the end of the spinney there was a terrible commotion going on. Blackbirds were shrieking out their alarm notes in top soprano. Thrushes were using the worst of language. Now they hurled forth their invectives from the top of the small pines. Now they left their high perch, and as they flew left a trail of shrieking fear behind them. Smaller birds, too, took up the panic note.

As I approached, the two owls sailed out of the wood once more. Evidently these were the bogies which had put "fear" into the hearts of the feathered world. As the two interlopers floated quickly out of the wood, I saw quite a number of birds plucking up sufficient courage to chase them off their preserves.

Then peace settled on the wood once more, and I left its glades to the trilling of the willow wren, and the soft cooing of the turtle-dove.

When we reached the hedgesparrow's nest, John was waiting for us. There on the ground, was one of the young hedgesparrows. It was moving slightly, and I stooped to replace it. "He's begun his cruel work already, ye see," said he.

John prevented me from putting the numbed youngster back again.

"No good," he said, "it'll be oot agin i' no time unless ye turn oot that bigger bird lying there wi' the other two."

I looked into the nest. How peaceful it looked, snug and warm, too. There was the young cuckoo breathing a trifle quickly, and two unhatched blue eggs. He looked anything but a cruel pirate ready to make his companions "walk the plank."

Then, as we waited, we saw the quiet of that nest begin to be disturbed. The interloper, the young cuckoo, began to fidget.

"He's quite blind," I said.

John nodded. "But look at 'im feeling aboot like wi' his wings. See! He's got one of them other eggs now atween his shoulder blades."

It was true. The little beggar had wriggled and wormed to some purpose, and now that the egg was riding on his back, he seemed to be resting from his efforts.

"Gettin' his breath for the final fling," said John. "Do ye see that holler in his back — that little cup into which that egg fits? It don't seem as though there is much 'chance' i' Natur' when ye see things like that, eh?"

I looked again at the drama of the nest.

"But that blind, helpless cuckoo can't possibly know that it *must* be the only occupant of the nest, otherwise it can't live?" I said to John.

"No," said he, "he's doin' it all by instinct. But what beats me is how that instinct grew; and why the cuckoo has such a fascination fer so many birds? Any amount on 'em 'll set to and feed it."

"What makes the young cuckoo start to wriggle and get the other on his back, I wonder?"

"They do say," said the keeper, "that after he's bin hatched a few hours, he grows sensitive to the others i' the nest."

"Do you mean that they irritate him?" I asked.

John nodded. "It seems as though he can't abear the feel on 'em — so oot they have to go."

I saw, as I turned my attention to the nest, that what John said was only too true.

The interloper was gathering his strength again. Now he slowly rose upon his legs. We hear about "special strength for special needs" — that is what I saw taking place before me.

As he rose, so he backed up the side of the nest with the egg, all unconscious of its fate, snugly ensconced in that fatal cup between the shoulder-blades.

Then, like Samson, as soon as he felt that the edge of the nest had been reached, he seemed to breathe forth his last prayer, "Only this once." A sudden frenzy took possession of him. Those little legs became empowered. They pushed upwards, outwards, until his back was an inclined plane, and with a final effort the egg slid and shot over the edge — into the cold and numbing death outside.

"I think I'll put it back," I said to John.

"And leave the cuckoo i' the nest?" he asked, with a smile.

"Then I'll take the young murderer out," I answered, feeling that the innocent ought to be protected.

"And will you kill the cuckoo in order to save the lives of the hedgesparrows? Better leave Natur' alone. What's wrong to you may be right to them becos it's nateral."

John and I left the nest, and we both for a time were very silent.

At last John said —

"If ye were to come back in a few hours' time ye would see that mother hedgesparrow sittin' on the young cuckoo, and sittin' there as unconsarned as though she hadn't a care i' the world. Right i' front of 'er would be her dead young uns. They might even wriggle, but she'd tak' no notice on 'em."

"She would not try to save their lives?" I asked, with wonder.

John shook his head. "She'd defend 'em whilst they were i' the nest, but her instinct don't reach beyond it, I reckon."

"I was going to ask you before," I said, "how the mother cuckoo gets her egg into the nest."

John paused a moment before he answered.

"That's a vexed question," said he. "I've seen the cuckoo meself carryin' an egg i' her mouth. I thought it were her own egg she were carryin' to drop into somebody's else's nest. But some think that she sits on their nest herself, and lays ordinary-like, just as other birds do."

"But what about the egg she was carrying?" I said.

"Oh," said he, "they say that it's one that she has taken oot o' t' nest that she's bin layin' in."

"Perhaps she does both," I said. "She may in some cases carry her own egg and drop it in, and lay it in the ordinary way in others."

"Well," said John reflectively, "I wouldn't put it beyond 'er."

CHAPTER SIXTEEN

The Otters' Landing Place

It was a gloriously hot day. I lounged under the shade of an old oak, whilst Raq worked his way in and out of a clump of bushes. I rather fancy there was a small rabbit right in the heart of the furze, but, do what the dog would, "bunny" wouldn't move, and the spikes of the gorse were deadly to his nose.

Then we heard the sound of footsteps, and a moment later Jerry came in sight.

"Where are you going?" I asked, beckoning him to a seat.

"Nowhere perticler," he said, with a smile, seating himself beside me.

"Then you're there already," I answered.

"What about a dip?" said the poacher. "It's the only cool spot to-day."

Raq and I got up at once and followed Jerry to the river.

Putting down my bag I noticed him staring at something on the opposite bank.

Here the stream had channelled its way out of stiff clay, and the red bank rose some seven or eight feet from the surface of the water.

I looked, and first of all saw a hole amongst the roots of a tree which grew on top of the bank.

"That's the hole of an otter," said Jerry, "and there's bin some young uns in it, I reckon. Ever seen one?"

I nodded and said, "Yes, once in the early morning when I was spinning for salmon, I saw coming towards me what at first I thought was a big fish cutting his way through the shallow waters. I stood still as a stone, and soon saw that it was an otter. It was like a big weasel, with a flatter head, and about the size of a terrier."

"What happened?" asked my companion.

"It came on till it was so near to me that I could see the whiskers each side of its mouth shining with the sparkling water that dripped from them. He was swimming like a porpoise, dipping alternately his head and tail. Then he paused for a moment, and with his bright bulging round eyes and big moustachios, he looked like Bairnsfather's picture of 'Ole Bill.'"

"That's him to a T," grinned Jerry. "A speakin' likeness, I reckon. But go on."

"Then," I continued, "he gave a big snuffle and —"

"That's when your scent struck his nose,"

"Dived," I continued, "and I never saw him again."

Jerry was silent for a moment, then once more called my attention to the hole in the bank.

"'E'd have an entrance like yon ower there, but under t' watter. Look at that hole there, there's a slide from it to t' watter below."

Then I saw that the clay bank was rough except just beneath the otter's entrance. Here there was a smooth, shiny track to the stream below.

"That's the otter's sledging hill," said Jerry. "They're playful critturs, and the old uns along wi' t' young uns slide full tilt from the hole to the river."

"It's like a toboggan or water-chute."

"I've heerd tell on 'em," said Jerry, "but I niver was on one. But when the otters tak' it into their 'eads to play, then the fun is fast and furious, and they plunge and frolic and slither into t' watter down that clay water-chute, as ye call it. You remember me pointin' oot a hedgesparrer turnin' ovver every little bit o' leaf and litter to find food?"

"The microscopic searcher," I said.

Jerry nodded. "Aye, if ye like. Well, the otter," he continued, "is just the opposite. When 'e's 'ard pushed, 'e goes far afield fer food, and nowt comes amiss. He'll tak' a young rabbit or water-hen, search a midden, or go doon to the sea and fill up on mussels. That's one o' the reasons why he's still wi' us, while his relative the pine-marten has a'most vanished. If that hedgesparrow uses a magnifying-glass to find its meals, then the otter uses a telescope, I reckon."

112

Raq got very excited when he saw me in the water, and jumped in after us.

When we reached the middle of the pool Jerry pointed to a ledge of rock rising just out of the water a few hundred yards from the "holt," which would have been difficult to see from any other vantage point.

"That's a reg'lar place fer that pair of otters to feed on after they've caught a salmon or a trout. Unless they've youngsters they don't tak' the fish to their 'holt.' They bring it 'ere. I've seen traces o' their meals there since I were a lad." We both held on to the ledge with our hands. "Have a look at the fish-scales left aboot," Jerry continued, pointing to the flat table-top.

"It seems worn smooth by the otters landing on it," I said, comparing it with other smaller ledges.

"Aye, I reckon otters 'as landed there fer 'underds o' years," said Jerry, as we had a sunbath. "The strange thing is that if ye could shoot or destroy all the otters on this river — clear t' whole lot on 'em oot — before long another generation ud tak' possession, and ye'd find they'd begin using this same ledge agin."

I tried to visualize something of what Jerry was telling me. I could see the ancient Britons spearing the salmon in the very pool at our feet. Fair-haired Saxons in their coracles used to glide down this river too. (A dark-haired child is rarely seen in the district to-day.) Up on the fells is a Roman encampment, and not far away is an Abbey, and

113

both the Romans and the monks liked the flavour of good fish. To-day the angler casts with his Hardy rod and Mallock reel, and his Harris tweed suit carries with it the odour of peat. Yet through all these changes the otter's landing-place has known no change.

"Ye're in a brown study," said Jerry, rousing me. "Ye've bin gazin' at t' river as though ye'd seen a ghost."

"I have been seeing ghosts," I said, and told him of my thoughts.

"If ye talk like that," he said with a smile, "ye'll give me t' creeps. I shan't want to come doon 'ere at nights alone no more — an' there's some good fish 'ere an' no mistake."

I laughed outright at the thought of Jerry being scared.

"I did git a queer turn once, though," he said, as he put on his clothes, "an' it were not far from 'ere neither. I were walkin' on a path by t' river in t' wood yonder aboot midnight an' —"

"Thinking of the ancient Britons, no doubt?" I interpolated.

"That or summat else," he added.

"With a steel 'gaff' tucked up your sleeve, you old rascal."

"Now ye're gittin' warmer. I'd only a fifteen-pun salmon fer company," he grinned. "It were in a sack ower me back. I must 'ave bin in a brown study, like you was, fer next thing I felt was a rush o' wings and a wallop on me 'ead that knocked me off me pins. I dropped t' sack and ran fer me life. I thowt Owd Nick were after me. Then I stopped an' laughed, as I 'eerd an owd owl scream oot agin at somethin' else goin' near 'er young uns. But I stopped laughing when I went back fer t' sack. It 'ad slipped doon t' river-bank, and t' salmon 'ad fell oot into t' watter."

"The moral is, 'Conscience does make cowards of us all,'" I said severely.

"I suppose," I said to him as we left the pool and made our way towards the loch, "that birds too are conservative in their love for certain places. Do those house-martins, Jerry, which nest under the eaves of your cottage, always return, or their young ones?"

"Aye," he answered, "they come reg'lar. There's a 'ole in a wall I knows of where fer ten years to my knowledge a pair o' wagtails 'as nested, and except fer one year they've 'ad a young cuckoo to bring up an' all. Then think 'ow t' rooks stick to t' same trees."

"That always makes me wonder. There are some not far from my house. 'Buses and traffic run right underneath their nests, and yet they come back every March. And you can't call rooks companionable birds, can you? For nine months in the year they're as wary as though you were their deadliest enemy."

We crossed over a small beck which was running its merry way to the river. The bank was riddled with the holes of the water-vole. Here and there were muddy places on the bank, just above the surface of the stream, so smooth that they might have been sand-papered.

"Even t' voles 'ave their fav'rite landin' places, ye see. In fact, the more ye see of animals and birds, the more ye see they're very like oorsels. Why, when YOU go walkin' oot, ye land up at t' farm or —"

"The old poacher's cottage," I chipped in.

"Aye," he said, with a smile which did me good to see. "Them's yer ledges where ye mak' fer and where ye feel at 'ome. Just as there's quiet places round 'ere where I allus likes to sit, and watch, and think."

CHAPTER
SEVENTEEN

The Despoiled Wood

Hearing the sound of men's voices whipping up their horses, Raq and I forded the river. To our dismay we found that they were busy cutting down a wood which was always a great delight to wander through.

What a shame it was to see those monarchs of the wood that had withstood the gales of years, being laid low. Ash, elm, larch, pine, sycamore, all were falling.

As I saw the havoc of the axe, an old melody came floating through my mind, and I felt I would like to plead that the wood might be left untouched. Alas! that a tree can be converted into pounds, shillings and pence! "Woodman, spare that tree."

I examined a few of the bigger ones. One old veteran had a fork near the top. Where the branches met, the junction made quite a deep well. In it was a nest I recognized. It was composed of twigs as a foundation, and then came thick layers of sheep's wool. This had been the snug abode of a carrion crow — a "doup" as they are called in this part of the country.

I should think that this fork had been used every year by these black marauders. The keenest-eyed game-keeper could never have seen a trace of it. No gun could have emptied its charge of shot into the young birds that had lain in it — the old birds being well able to protect themselves. Here also was an oak. In a neat hollow where the branches divided, lay the snug nest of an owl. For how many years, as dusk fell, he had hooted his terrifying notes to the fields that lay across the river — a bogey to all mice and voles — no one would know. Now the trees are laid low, and somewhere there may be birds who mourn along with me that these shapely trees no longer sway at the wind's caress.

Down by the riverside I found a wooden shanty. Underneath it a steam-saw buzzed like some venomous bee. Here the logs were being sawn into required lengths. How fragrant and resinous smelled the pine!

I strolled up to the man who was working the machine.

"What's that you are sawing up?" I asked.

"Elm," he answered. "It's goin' to be made into table-legs."

"And the pine?" I queried.

"'Chocks' for the coal-mines," he replied.

"Is there much beech?" I continued.

"Aye," he said, pointing to the topmost fringe of the wood, "there's a goodish bit up yonder. They'll go to the makin' o' bobbins for t' mills."

"There are not many ash-trees felled," I said.

"No," he answered, "the sap's still runnin' in them. It doesn't matter so much with the elm and the larch. If we cut the ash down now, it won't keep so well. We're leavin' 'em till later in the year."

As I left the slaughter of the wood I felt sorry for all the trees. It has taken a century to grow that glorious wood. Two years will see it wounded and naked. I felt most of all for the beech-trees. I have sat underneath them and marvelled at their sleek, well-groomed trunks. I have seen the shimmer of moonlight on their silvery bark, and heard the wood-pigeons coo from their outstretched arms. Soon they will be in some mill, deafened by the sound of a hundred looms. What a transformation! Up through their centre runs a core of darkened hard wood. The woodman had referred to this as the "heart" — the lighter wood surrounding it as the "sap." I wondered, as I saw them waiting for their execution, whether the tree had a soul. As I stand underneath them, I sometimes think they have. Rustle, whisper, quiver — how sensitive they are to Nature's caresses or blows!

For a few moments I watched the mighty Clydesdales heave and strain as they pulled the tree corpses to the saw. It was a good sight to see their shining coats and watch their mighty sinews grow taut at the pull of the load. It is hard to believe that such noble creatures started life as quite insignificant animals. Millions of years ago, so the fossil remains tell us, the ancestors of our modern Clydesdales and Shires were no bigger than a collie dog and possessed five separate toes. They roamed over the wide world.

Other remains show that during the centuries that followed, the wild horse increased in power and size, and lost one of its five toes. During the ensuing years the primeval horse went on growing and yet losing its toes. Eventually one grew bigger than all the others and the horse walked on this alone. The others had dwindled into nothingness. To-day, to all appearances, the horse has only one solid foot encased in horn. Yet the remnants of the five toes can still be found.

Noble beasts they are, with their sensitive ears turning backwards, forwards, sidewards, and a muzzle softer than the smoothest velvet. Motors? Dirty, oily, mucky — a soulless nexus of cranks, shafts, pistons, and what not. Give me a horse!

I was walking home feeling rather sad about it all when we met Ned.

I knew that he of all men would understand.

"Don't ye worry, Natur'll soon clothe it agin," he said. Not being so optimistic, I didn't reply.

At the next cross-roads we came to a patch of land which had been used as some sort of a rubbish-tip. I remembered it when it lay unsightly with cinders and broken earthenware. Now it was a waving mass of green nettles. Amongst them were those which had a white bloom.

"That were an eyesore not very long ago," said Ned, "and Natur' hates an ugly patch. Give 'er a year or two, an' she'll allus drape it with beauty. There seems to be a con-con —" Here he halted in search of the particular word he was thinking of, but which eluded him.

"Conspiracy," I suggested.

"Aye, a sort o' conspiracy i' Natur' to cover up ugliness. Look at that old cottage ovver yonder. It's bin left to tumble doon. It were a blot on t' landscape. But already the mosses and the ferns and the grasses are creepin' into its hollers, and the stone-crop is beginnin' to tak' possession — it'll be a lovely old ruin i' a few 'ears."

I did not interrupt him, though I wanted to ask him a few questions. But I could feel that he was delighting in giving play to his inmost thoughts.

"Aye," he said, "as soon as human bein's left that cottage, then the winds blew into its cracks sleepin' life. Birds and beasts carried seeds and dropped 'em on its bare walls, and the evenin' whispers o' the breeze slipped 'em into little crannies where some bit o' earth lodged."

"There's that wood ovver yonder too," he continued.

"Black Wood, wasn't it?" I asked, looking at what had once been a miniature forest but now lay almost denuded of trees.

"Aye, Black Wood it were called, and Black Wood it were," said Ned; "a dark spinney as ever ye cam'

121

across, wi' nowt but shadders and tangled undergrowth to fill it."

"It's a mass of bluebells now," I said.

"Aye, that's Natur's way," he said. "Let the light into a place, and the wood-sorrel and the primroses'll soon smile at ye. Ye can shut oot the sunshine fer 'ears, but Natur' bears no grudge. She's ready to get on wi' her job o' makin' all things beautiful. So don't ye worry. Natur' will use all her beauty-aids. The winds, the streams, the birds'll carry seeds to all the wounded acres."

He paused a moment and then said quietly, "And the wilderness shall blossom as a rose."

CHAPTER
EIGHTEEN

The Jewel of the Stream

It was a beautiful morning when I walked into John Rubb's shop. Why I went in who can say? Perhaps it was because I had overheard a chance remark that the river had received an influx of fresh water. Or perchance the thrush which had sung persistently outside my study window had rather unsettled me for serious work.

When I entered the shop, John was himself attending to an old farmer, whilst his assistants dealt with other customers. This is my friend's forte. These ploughers of the remote solitudes do not like shops, but they like John. In his hands they do not know they are buying anything, they feel they are "making a deal." As the old man went out wearing his purchase, my friend, looking sideways at me, said — "And what can I do for you this morning?"

For answer I glanced out of the window. John, after a slight pause, said — "I have to go out to see a customer this morning at Bridebridge — a bit of business I want to get through. Care to come?"

And that is how we came to find ourselves in the old Ford packed with rods, lines and tackle, with Raq sitting up in the back seat, his pink tongue lolling out for sheer joy and his great brown eyes laughing with soulful delight.

I should like every one to see Bridebridge. The only thing alive is its river and the feathered folk that flit on its banks. The inhabitants of that tiny hamlet are somnambulists, and will never wake up in this world.

But the river is Paradise. It steals down through giant crags which it in vain has tried to wear out. Even when the flood hurls itself at their mighty bastions, they look on, serene and unperturbed. Then, as though worn out by its exertions, it glides gently through quiet glades where the alders use it as a mirror, and the shingle transforms its anger into laughing ripples.

A little later John and I parted, each of us setting out to fish his own particular part of the river.

The water was in the state known as "clearing off." It was a clear rich brown, and it was not long before I had creeled a couple of beauties.

It is not merely the catching of fish which brings joy to the real angler's heart — it is the things he sees and hears.

After I had been fishing for some time, my eye caught the gleam of a bright jewel as it flashed down stream.

"Kingfisher," I said to myself, and going very warily round the bend of the river, screening myself in the bushes, I saw three youngsters, each one an emerald and ruby gem, seated on a stiff root which jutted out from the overhanging bank.

I quietly put down my rod, and wished John were with me to see them.

As I was wondering where he might be, a wild duck came flying overhead, following the course of the waters. He did not see me, as the bushes veiled me from his alert eyes. But as he sped down the river, I saw him throw himself suddenly higher in the air. I knew then where my friend was.

In a few minutes I had reached him, and after telling him of the young kingfishers, we both crept back to a hiding-place from which we could view them.

John touched my arm and pointed further down the river. There, over a quiet backwater, sat the parent bird. From time to time he turned first one eye and then the other on the water beneath him. Then there was a downward plunge — he dropped like a stone, to emerge with a minnow in his bill. The water dropped from off his brilliant back like shimmering diamonds as he alighted on another perch and hammered the fish to death, prior to feeding the hungry nursery that waited impatiently for his return.

"That bird," said John, "is really a land bird, like the dipper I told you about, but it has changed its habits and environment. It can't stay under the water long, like the dipper, you notice. Just in and out — that's his method of fishing."

125

"Where's its nest?" I whispered.

"Some hole or other not very far away. Look," he added, "there's the other older bird feeding them now.

Across the water another kingfisher, as brilliant as the first, was attending to the needs of the fledglings.

"Which is the cock of those two?" I asked.

John shook his head. "Now you're asking me a question. I can't say; they are so much alike."

And so for a few minutes more we had the joy of watching them. As we moved away to have our lunch, John said — "It's a curious thing that the most beautiful birds come out of the dirtiest nest."

As we munched our sandwiches, I asked John if he had ever seen a kingfisher build its nest.

He nodded.

"They select their site in some overhanging bank, and then fly full tilt at it with their sharp bill," he said.

"That won't make much impression on a hard bank," I said.

"They stick at it though," said he, "and after a few attacks make a hole big enough for their feet to hold on to. Then it's navvying work until they've channelled up a yard or so."

"And the nest?" I asked.

"A stinking mess o' fish-bones and litter," said John, "and out of it comes that necklet of beauties we saw up yonder."

126

For a moment or two my mind was busy recollecting whether I had ever come across any other dirty nests. There was the chaffinch's — clean and tidy. I thought, too, of thrushes and waterhens — these were quite sanitary. Then I remembered the gannets on the cliffs of Ailsa Craig, how filthy were their nurseries — fish-eaters too. I was recalled by John saying — "I've noticed too that when birds nest in a hole and are free from observation, that the hen bird decks herself out as brilliantly as her mate."

"You mean that there is no need for protective colouration?" I asked, and I told him what John Fell had told me about plumage.

John nodded. "Think of the shelduck," he said, "a magnificent beauty in chestnut, black and scarlet, and ivory. She nests in a rabbit burrow where no colour can attract a marauder. You've often seen her when we've been together on the salt marshes. She doesn't need to go about in a drab dress like her cousin the mallard."

"What about the mallard drake?" I asked. "He's brilliant enough in the breeding season."

"True," said John, "but he doesn't hang round the eggs and call unwelcome attentions to them. He keeps himself apart from such worldly cares. And talking about colour," continued John, "you should see the eggs of the kingfisher. They are a lovely white. so white and transparent that they seem to gleam in the dark where they are laid."

John pointed out to me, a little later, the hole where the young kingfishers had first seen the light.

In my mind's eye I could see the long corridor running up into the darkness. Then I saw the hen bird return to the nest, alight, and walk up that dark corridor. When she reached her treasures, she knew exactly where they were, for every white egg snared the twilight which filtered into the dark chamber and reflected it — little round stars shining in the dusk.

John and I returned to our fishing, and as usual it was not long before we had netted some good fish. Suddenly, however, the peace of the sanctuary was broken by the distant baying of dogs. Raq pricked up his ears. Over the quiet fields there came blood-curdling baying.

"Otter hounds," I said.

Nearer and nearer came the uncanny baying. Then a score of rough-coated hounds came in view — something between a foxhound and an airedale.

Noses were down on the bank, whilst tails waved aloft with joyful and good-humoured pliancy.

Then came the hunters and their usual crowd of followers. Gallant fellows all were these seekers of the otter, dressed in royal blue, with hefty staves in hands. Ladies, too, followed those waving tails, not on horse-back, but on shank's pony. Otter-hunting is hard-slogging over the roughest of country. Streams must be forded, and high wooded banks negotiated, if a glimpse of the quarry is to be obtained.

Soon the deep baying of a hound told all that he had picked up a scent. Other dogs ran to the spot, picked up the trail, and a full-throated chorus, together with a mad rush along the banks of the stream, announced to the hunters that the otter might soon be found.

A mile or more the dogs carried on, then spread themselves under the roots of tree and bush that shadowed a big pool. Somewhere between those rocks, or perchance under a hollow tree lurked the otter.

He had done no harm, save perchance that he had a taste for salmon and trout. But he was there to provide sport for huntsmen, who in ordinary life are as tender-hearted as most people.

Naturally the otter sought the seclusion of his hole in the rocks, but a keen terrier soon scented him. So once more he was in his natural element, the water, making for the head of the pool. All the time the baying of those terrible hounds sounded in his ears.

But alas! at thc head of the pool he found a string of men standing in the shallows, so once more the poor beast turned to face the baying of the dogs.

Like a wraith he slipped under water, and made for the tail of the pool so that he could glide into security further down the stream. But once more he found a chain of hunters preventing his exit.

He was becoming weary now — weary with the long run up stream — weary, and perhaps wounded from the bites of the terriers. But there was no friendly "holt" into which he could creep, safe from his pursuers. So he glided under the bank where the

129

water was deep, and the shadows mercifully wide. Here he sank his body, leaving only the tip of his nose above the surface — enough to breathe. But that tell-tale nose was espied by some keen-eyed huntsman, who prodded with his pole. Once more the harried, jaded brute had to face those grim, baying dogs, and those merciless men.

And so the cruel business went on until, with breath spent, and all holes barred against him, his furry body was at last gripped in the jaws of a hound. It was soon over, and the less said about "the kill" the better. Nothing was left save the paws, which went to decorate some huntsman's cap or perchance was worn as a brooch by some fair member of the hunt.

"It's been good sport," says one follower to another.

"Great fun," he replies.

But John and I walked back to the car in silence. There was no need to express our feelings.

CHAPTER
NINETEEN

Keeper and Poacher

"The birds are gathering into flocks," said John Fell as Raq and I walked with him across the fields. "That's a sure sign that sunny days are comin' to an end."

We were tramping through a stubble field and he pointed to the numbers of small birds of the finch variety that took refuge in the hedges at our approach. "Linnets, goldfinches, chaffinches, most on 'em are," said he. We saw them descend to feed on the seeds — the goldfinches looking like drops of sparkling sunshine and red fire as they searched the towsled heads of the thistles.

"They are not communal birds then?" I asked.

The gamekeeper shook his head. "The nestin' season finds 'em living their own little lives wi' their mates. 'Each family fer itsel'' is their motter. But when the autumn begins to touch every leaf wi' yaller and scarlet, then they join up i' bands, and feed together through the winter."

For a moment or two we watched the little companies. Not for long did any of them settle on the ground. Each moment some alarm or menace

131

would send them like a fluttering cascade to the safety of the hollies and hawthorns. Many of them swayed from the topmost branches, where they could be seen surveying the land for the peril that lurked in the shadows.

"Jerry was telling me," I said to John, "that many birds and beasts are now scattering their families right and left, driving them away from the old home, and sending them out into the wide world to seek their own fortunes."

The keeper nodded. "That's true," said he. "Fer some time I've heerd the owls kickin' up an awful shindy. They're drivin' the youngsters away in order to find fresh huntin' grounds fer theirsel's. Food'll grow scarcer as the days shorten. Jerry's right, I reckon."

"Then," I persisted, "why do these birds" — here I pointed to scores of finches on the stubble — "do the exact opposite? More mouths mean less food for each."

"Well," said he, "I reckon they do it partly fer protection. They've got many enemies on the look oot fer 'em, and they know it. That's the meanin' o' their continual swinging up and doon from hedge to stubble. And when there's a score or two together, it means that there's scores o' eyes and ears on the look-oot, instead o' just one pair. Also, I fancy, that though there are more mouths to fill, yet when they're i' flocks there's more eyes to seek and find — and it's mostly the seed-eaters that gather inter packs."

132

He opened a gate that led into a field which was dotted with numerous bare bushes.

"Mice, which owls live mostly on, are not as numerous as seeds, and so them old 'screechers' have to consarve their supplies by gettin' rid o' youngsters. If ye tak' perticler notice, ye'll see that all the hen chaffinches live by themselves, and all the cocks seek out each other."

"And why do they do that?" I asked.

"Can't say," said John. "P'raps they think that absence mak's the heart grow fonder. Or p'raps they've got tired o' each other's company and want a change. I sometimes think that it wouldn't be a bad plan if we men-folk —"

Then he stopped, and I could see the twinkle dancing at the back of his eyes. But he never finished his sentence.

"This is what I came to 'ave a look at," said he, pointing to the field which stretched out before us. Raq was already investigating it, and was paying particular attention to a small patch of ground in the centre.

"That's where the partridges jugged last night," said John, watching the dog, who was puzzled by the fact, that though the scent was strong on their camping site, yet he could find no trail leading from it.

"That's 'cos they flew right up and were away wi'oot touchin' t' ground, old man," said John, as he came up to where the dog was casting round for a scent.

"Those bushes," I said to him, "are to prevent poachers, aren't they?"

"Aye," replied he; "I put 'em in as soon as ever harvest is ovver. If I didn't, a couple o' poachers wi' a good net could sweep the field and bag a whole covey o' partridges as were sleepin' there — they lie so snug an' close."

I went up to one of the bushes. "You don't dig it in then, and fix it firm in the ground?"

The keeper shook his head. "It gives more trouble to them gentlemen, I reckon, if left loose. Them thorn bushes get more entangled i' the nets — leastways, that's my experience."

I looked round the field. "You don't seem to have any regular formation in the placing of the bushes," I said; "they are not spaced out regularly."

"There's a bit o' method, though, i' my madness," John answered with a smile. "If ye'll look roond ye'll see that they're put i' places that are opposite gaps where night prowlers are most likely to come inter t' field. There's that low railin' ovver there, a very likely place fer them night 'awks to pop ovver — well, there's some good spiny hedgehogs waitin' to give the net a welcome. That's the principle I've worked on."

Once again I looked over the field, and saw that all the weak places were well guarded.

"Generally speaking," I said, "you do not seem to bother much with the parts lying near the hedge."

John shook his head. "Partridges generally snuggle together right i' the middle," he said.

The next field had plenty of natural cover, and Raq was having a great time nosing the rabbits from their seats in the coarse grass and giving chase. Once he got so near to one that his yelp split the air.

"He doesn't often 'give mouth,'" said the keeper, "but ye wouldn't have to let him chase rabbits like that if ye used him wi' a gun. If he saw any game in front o' ye, he'd break away and run in and spoil your shot. But ye never use a gun now, do ye?"

I shook my head. "Never again," I answered. Seeing scores of white tails bobbing into the hedge bank, I asked, "And why don't you bush this field? There are rabbit poachers as well as game poachers, aren't there?"

John paused to light his pipe. "That's true," he said, "but it wouldn't be much good bushin' a field to stop rabbit poachin', fer their manner o' usin' the net is different, and ye've got to tak' into accoont that they're after rabbits — which 'as very different habits."

"Ye see," said John, pointing to the warren, "when the rabbits first come oot towards dusk they don't run very far away from their burrers. Ye can see how short the grass is cropped near their holes. Then they git more confidence, and want to git oot to the better feed in the middle o' t' field."

"I can see the lanes they have made in the grass," I said, tracing some of the well-defined pathways that crossed and criss-crossed in front of me.

"Well," continued John, "by ten or eleven o'clock at night all the rabbits are well oot i' the middle there. But it ud be no use fer them fellers to sweep the field as they do when they want to snare the partridge — there's a much better way. Onlike the birds, the rabbits have a definite place o' refuge to mak' fer if disturbed. So the poachers simply stretch their net a few yards from the burrows, parallel to the hedge. Then, when all's ready, they give the word to their dog, and, circlin' the field, the lurcher soon sets the rabbits movin' fer their holes. Scentin' the dog, they race fer their homes, only to dash headlong inter the net. The poachers keep behind the net and move quietly and quickly along as they feel or hear a bunny in the toils."

John then picked up an imaginary rabbit by its hind legs and brought his hand down smartly on the back of the neck, indicating the short shrift that the entangled animal received. "Then," he added, "inter the sack it goes."

He shook his head. "No — bushin' a field ud be no good fer such work as that."

As John was in a hurry we returned home by the road, instead of the fields, much to Raq's disgust. It was not without interest, for the first thing he found was the poor crushed body of a hedgehog. It was amusing to see how gingerly he sniffed at the carcas, as though he were not sure that even in death the spines might not trouble his nose. Once, in a fit of exuberance, he had touched a hedgehog — but his nose was sore for nearly a week afterwards, and Raq never forgets such experiences.

"Killed by a car," said John disgustedly.

"I have seen quite a number on the roads," said I. "They seem to be getting more numerous."

John shook his head. "It isn't that," said he. "That poor little beggar were trekkin'. I' one way he were migratin' like the birds."

"Migrating?" I asked with interest.

"Well," said the keeper, "he's bin turned oot o' his 'ome. He were born aboot July, and he'd 'ave six or seven brothers and sisters."

"It cannot be particularly comfortable to lie beside an inverted pincushion."

John laughed. "Oh, that's all bin thought oot all right," said he. "When they are born, they come into t' world wi' soft spines — they're not a nest o' young furze-bushes."

"Then," continued the keeper, "they've bin looked after wi' great care till just aboot now, when the owd uns know that their bit o' territory 'll feed two on 'em but won't be able to support a big family — so they clear the youngsters oot, with just one thought in their little brains, an' that is to be sure and roll up into a ball when danger threatens. 'Curl up, curl up,' grunts the mother when any sound comes along that her ears can't understand, or her nose can't explain — that's dinned into their ears all day long."

"So you think he was just setting out to find pastures new when sudden death struck him? Poor little fellow."

"Aye," said John, "by bad luck he struck the main road soon after leavin' home, and found it easy to travel on. He shuffled along wi'oot meetin' any adventures. P'raps he thought he had found a fine place fer food, fer a lot o' dead things lie on t' road. Then he heard t' sound of the comin' o' a motor and felt t' shake o' t' ground as something rushed towards him. He didn't know what it were, but he remembered his mother's 'Curl up, curl up.' So he curled up, till speedy death uncurled him for ever."

"Let's give him a decent burial, John," I said.

So we found a rabbit burrow in a quiet spot and placed the little fellow inside, much to Raq's disgust.

CHAPTER
TWENTY

The Harvest Field

"Where shall I find them?" I asked Sally, Joe's wife, as Raq and I entered the farmyard.

"Joe is with his sheep. Alan's somewhere about the buildings."

Thus is the whole farm divided, the house, the fields, the buildings.

"I'll find him," I replied, glad of the chance of exploring sheds and byres, granary and barn. I delight in the aesthetic beauty of stately buildings. Cathedrals and mansions produce wonder and delight in me. But for sheer unadulterated pleasure let me wander through what the farm knows as "its buildings." Let me smell the mixture of horsiness, fodder, and leather harness of the stables, the warm, sweet smell of cattle waiting to be milked, the juicy freshness of turnips heaped before the big chopper, and, as an accompaniment, let me hear the call of the calves for their breakfast, the satisfied grunt of the old sow as she pushes her nose deep into the food trough, and the call of the hen to her chicks as she finds some dainty morsel for them to share.

Alan was just coming out of the stable.

So, with the Clydesdale walking at our side, I accompanied him to the High Barn, where the last of the cutting was to be done.

Some of the field was already in stook. From the commanding hill I could see distant fields, many of them shimmering with golden stubble, others ready for "leading."

"All the fields seem to be set out in the same order," I said, pointing to them.

"Aye," said he, "we tak' aboot eight sheaves and stand 'em up, north to south."

"Why north to south?" I asked.

"So as to get all the sunshine they can," he answered in a tone which had a soupçon of pity in it for one who was town-bred.

Soon the binder was circling the remaining oats. I watched it with fascinated eyes. The slain stems fell on to the canvas elevator and were gathered into a sheaf. Then mechanical arms encircled it with band (string), and mechanical fingers deftly tied a knot. The next moment the sheaf was spurned and cast out on to the field. There was an uncanny heartless precision about it that somewhere in my inner self raised rebellion against its maddening accuracy and relentlessness — something that seemed to sum up the very spirit of the city life in which I dwelt.

"Different from the old days, eh?" said Alan, "when the men and women had to select their own bundles and mak' their own bindin's and twist 'em into a knot, eh?"

"There's a knack in making that knot, too," I said. "I've tried many a time, but mine would never stand much knocking about."

The "ten o'clock" in the fields is always a pleasurable interlude. Standing on the hill which overlooks the farm, one sees Hannah sally forth from the kitchen door carrying in one hand a large can of tea and in the other a basket covered with a white cloth.

As soon as her figure is seen there is a visible relaxation amongst all the workers, and the keenness for work somehow loses its edge. Finally all repair to the shelter of the stone wall.

The view from the High Barn is always one that impresses me. In autumn it is specially beautiful. Below stretches a valley of green and gold. From quiet farmsteads the blue smoke curls upwards. Here and there a wall of a house is flushed crimson, its virginia-creeper dyed deeper than a setting sun. Behind me the fells lie with folded hands, whilst on their breasts glow purple acres. Grey rock peeps out from shadowed ghyll and the high foreheads of their summits. From the open spaces around and below me come the cawing of rooks, the staccato cheekiness of jackdaws, the lowing of contented cattle, the whirr of wings, as companies of sparrows and finches make the most of what food still remains.

By my side Raq sat watching all of us intently, ever ready for the piece of crust or scone which he knew would be tossed to him.

After the pipes had been lit, Alan said, "There's summat ower yonder for you to look at."

We walked together towards the top of the field, where he pointed to a sheaf standing by itself. Here we saw intertwined in some of its stems a small grass nest about the size of an orange.

"We saw it just afore the binder got to it," said Joe, "but we've seen nowt come oot on't. What is't?"

"Well," I said, picking up the sheaf, "I think it's the nest of a harvest mouse."

"The same as ye find in the stacks i' winter-time?" asked Alan, and I thought there was a note of disappointment in his question.

"Not exactly," said I; "I've only seen a few before, and that was when I was farther south. It's the first one that I've come across in this district."

"It's a very small nest," said Joe.

"And there's no entrance," said Alan. "How does he get in and out?"

It was not until Alan pointed this out that I noticed that the outside covering of the nest, which was very neatly woven, was unbroken.

"Nobody really knows just how he makes it like that. I have been told that one mouse stops inside while its mate brings it the grass to work with. It's only a couple of inches long, and it's got a tail just as long. It's a pretty sight to see the little beggar run up a stem of wheat, and nibble at the ear."

"And where do they go in winter?" asked Joe.

"Some of them," I said, "go into their burrows in the fields, and sleep like the dormouse, though

142

perhaps not so heavily, and some of them get taken up in the sheaves and find themselves in a stack. Then they have the winter of their lives feasting on the good things there . . ."

"Aye," said Alan, "ye should have been here last year when we were threshing. There were scores o' mice and a goodish few rats, and there were three or four owls screechin' and hollerin' round that Dutch Barn, and what a feast of mice they had. What beats me is how they get to know."

"What tells the carrion crow there's a carcas? How do stoats and weasels learn that a certain district is over-run with rats?"

"Well, them owls got to know anyway," said Joe. "T' mice were rustlin' aboot in the loose chaff, and some on 'em were runnin' up t' beams which 'old up t' roof. With the sheaves all gone, the little beggars 'ad lost their runs and didn't know where they was goin'. Moon were shinin', and every now and then along came the owls. I could see 'em hover fer a minute, give a scream which must 'ave sent a shiver doon the spine of every crouchin' mouse, then doon they dropped on t' chaff. By the heavy way i' which they rose, I knew they'd got summat i' their talons. Now and then they checked theirsen as they were flyin', and picked 'em off t' beams."

"And what did they do with them?" I asked. "Eat them on the spot?"

"No," said Alan; "they grabbed 'em, and then carried 'em off to their larder i' that wood by t' quarry. They'd 'ave a fine cupboard full by mornin',

I reckon. They haunted that Dutch Barn till there weren't a mouse to be seen. I wouldn't care if they'd only let chickens alone. Lots have had their heads pulled off."

"Hard luck, but owls really come to get the mice. The mice would find easy food left about from what Sally with too generous a hand had given the chickens, and then the owls turned their attention to the chickens. But they prefer mice first every time."

"It sounds sense," said Joe as they returned to work.

The wild things that had found sanctuary in the standing oats could now be seen dashing madly about in all that remained. Here and there little brown heads, with staring, fearing eyes, peeped furtively out, when the rattle of the binder had passed.

From another part of the lessening square of oats there came a shout, and men dashed after something that ran, something that bumped into the fallen sheaves, and, unable to find a clear path to the hedge, crouched and waited for the death which was sure to come.

"There are only a few left in now," said Alan, pointing to the dead rabbit that swung from the swarthy arm of one of the men. "Most on 'em cleared oot durin' the night, I reckon."

Raq proudly brought one to me which he had caught. I held the poor little chap carefully in my hands remembering the ripping power which lay in its strong hind claws. But I had not the heart to kill it, so waiting my opportunity, I walked over to the hedge and dropped it down a burrow. Raq looked at me with reproachful eyes.

It was not long before the field was finished and the men had set it up in stooks.

"That's the last of it," said Alan, as we left the field to go homewards.

As I turned to look at the golden field in which I had spent many happy hours, a lark rose and, sailing over it, sprayed it with melody.

"Not quite so joyous as in the spring," I said, as we listened to its music.

"It were allus a great place fer larks," said Alan.

I watched the bird, no longer flying into the eye of the sun, but rather in an arc as though surveying the whole field. I fancied that it was singing the "swan song" of its old sanctuary — a hiding-place no longer. How many times the oats had saved it from the sparrow-hawk's dash none would ever know.

It was dusk when we returned to the farm. This is the hour above all others when everything speaks of restfulness.

One by one the hens walk sedately towards their roosts. As they fly up to a chosen place, there is the sound of a scuffle, and disturbed ones scold the interrupter of their peace. Then quiet reigns.

Across the yard waddles a string of ducks, giving spasmodic quacks, lazy cackles, ere they snuggle together for warmth in the corner of the duck-house.

A calf "moss" with a low resonant note, a cow in the distance answers with the tone of tender solicitude; the horses crunch — crunch — crunch in the neighbouring stable; an old sow is breathing heavily in her sty.

Overhead, little pipistrelle bats twist, wheel, turn — black elves in an eerie dance. Trees, buildings, fields, slowly merge themselves into a mass of velvet shadow. On the darkening flags which surround the back of the house, the yellow light from the kitchen lamp glows warm and inviting.

Then for a moment the hush is broken by the men cleaning their boots with the stiff yard brush which stands by the door. There is the harsh rattle of a metal door-latch, the last stamp of feet on the flags — the day's work is done.

CHAPTER
TWENTY-ONE

The Hare's Leap

"This is the first touch of hard frost," I said, as the snow crackled under our feet as Jerry and I strode over the fields.

Before us the countryside was a picture of beauty. Frost, that wondrous lace-maker, had been at work, and every twig and blade had been outlined by icy fingers. There had been a fall of snow. It had settled on the ground like a white powder rather than in flakes, and every branch was a study in black and white. The sun was veiled so that it looked like a ruddy moon.

"There seem to be millions of rabbits about," I said, as we walked by the fringe of a wood and noticed the marks left by their feet.

"Aye," said my companion, "a light fall seems to stir 'em up a bit and mak's 'em scamper aboot more'n usual. Ye can see by the way that their tracks run i' circles that they're playful critturs, allus ready fer a game o' 'chase me chase.'"

"And in heavy snow?" I asked.

"All of us have a bogey, and I reckon that snow is the bogey of everything that flies and runs. Clean hard frost they can face fairly well — but snow not only covers up their food supplies, but mak's 'em leave their tell-tale tracks fer all enemies to see."

Everywhere in the white fields we could see confirmation of Jerry's statement. Here, the rabbits had run round in aimless circles. There, some of them, struck by a sudden inspiration, had begun to scratch down to the hidden herbage. But most of them had seemed to be bewildered. To scores it must have been their first experience of the white terror.

A joyful yelp called our attention to Raq.

"He's put a hare up," said Jerry, "and must have got pretty close up to it fer 'im to have yelped like that."

We ran towards the place where we had caught sight of the vanishing dog. Right through the hedge the tracks ran. Then in the next field we saw Raq lumbering along with his nose on the ground and his big ears flapping like two large fans.

As for the hare, she was sitting at the top of the next field, shaking the snow off her pads and apparently quite unconcerned that the spaniel was on her track.

"She's got his measure o' speed all right," said Jerry, laughing. "She knows he's got as much chance of ovvertakin' her as a tank 'as of ketchin' a racin' car. Ye'd better whistle 'im off I'm thinkin'."

So I put my fingers to my mouth and blew shrilly for the dog to return. I knew he heard me, for I could see him falter in his stride. For a moment love of the chase strove against the discipline of many years. Then obedience won the day, and he came back with his tongue lolling out of the side of his jaws, and a rather apologetic look in his lustrous brown eyes.

"That's where she were sittin' ye see," said Jerry, pointing to a snug seat hollowed out in the stubble and quite open to rain or snow.

"There's not much shelter there," I said. "You would think that she would at least choose a place that had some covering over it in case there was a heavy storm."

Jerry did not say anything for a moment, but followed the tracks which the dog and hare had left behind.

"I've known 'em let the snow cover their bodies up completely," he said on returning. "There must 'ave bin two or three inches lyin' on their backs. O' course, they fit into their 'form' just like a finger slippin' into a glove, and then t' snow acts just like a roof."

"I should have thought that they would have been smothered."

Jerry shook his head. "Their warm breath thaws a kind o' chimley, and they lie there as snug as a bug in a rug. Did ye notice which direction that owd hare led the dog?"

"Right up that hill over there," I answered, pointing to the place where she had waited.

"Why didn't she lead 'im doon towards t' river yonder? Why uphill?"

I looked at the dog whose breath still came in quick pants.

"To make him short-winded," I answered, "and give him a stitch in his side."

"Mebbe you're right," answered my friend, repeating what John had told me, "but she can run quicker uphill than doon — her front legs are fairly short, and her back uns are long, and as she runs uphill her hind legs prop 'er up and keep her fairly well on t' level, ye see."

We followed her tracks until they reached the hedge.

"She stopped 'ere fer a second afore she leapt through, ye see. Hares allus do that. That's 'ow them nasty poacher fellers (here he winked slyly at me) can allus git an easy shot. A rabbit goes clean through a hedge wi'oot thinkin' what's on t' other side."

We scrambled through a hole in the bottom of the dyke.

"I was just thinking that you know too much, Jerry," I said severely.

"Now ye can see how she's got ahead o' t' dog. Look at the marks of 'er hind legs — them long uns there. We'll measure how far it is to the next — they're as clear as daylight in t' snow."

Jerry stooped down and began to span out the distance between the leaps.

150

"First leap after comin' through the hedge," he said, "were aboot eighty-six inches."

"About seven feet," I said in amazement.

"The next un is ninety-five — she's puttin' a spurt on. The next drops doon to forty. She's away agin now," said he, still following and measuring the space between the leaps. "This un goes up to aboot a hunderd; now she drops agin to forty-two."

"She seems to take a big leap and a small one alternately," said I.

"It looks like it, though I never noticed it so clearly afore," said Jerry; "and here's biggest o' t' lot — a hunderd and fifteen inches."

"Nearly ten feet," I said, looking in wonder. "No wonder that she left you far behind, old man," I said to the dog.

"When a hare leaves her young uns — ye know she puts each little un in a diff'rent place and —"

"Is it so that a stoat can't bag the lot?" I asked.

"Aye. But when she gits near one on 'em, she'll tak' whackin' big leaps at right angles to her real direction. This breaks 'er scent, ye see. One minute a stoat has her tracks at the end o' his nose, and next minute it's gone — vanished into thin air."

"You wouldn't think that a leap of ten or twelve feet would make much difference to a hunting stoat," I said.

"Ye forget that ye're a great big chap o' six feet that can see ower a wide stretch o' country. A stoat's only a li'l feller, near to t' ground, and if he does pick up the scent agin he soon comes to another

151

blank — it's enough to break the 'eart even of a stoat."

As we walked on by the river Jerry pointed to a track left by some three-toed bird. He looked at me.

"Heron?" I said, drawing a bow at a venture.

"No, water-hen. The way she walks ought to tell ye summat about 'er ways."

"Well," I said, "those splayed feet mean that the bird can walk well on mud and weeds, and the tracks are in a straight line because —"

Here I faltered, and Jerry came to my rescue.

"Because," he said, "the water-hen 'as to move quickly through tangled weeds, so cuts its way through with one foot, and the other puts itself doon i' the same line. So it cuts the narrowest track fer itself in the tangled herbage."

"It has no webs like the mallard," I said, looking at the imprints again.

"No," said Jerry, "that's why it is such a clumsy swimmer. A duck seems to glide along. But a water-hen constantly keeps a pushin' its neck and head forrard as though it didn't get enough push from its long legs."

Across the brook there lay the trunk of a fallen tree. On this were to be seen small separate foot-prints. So clear and neat were they that the four toe-marks of the fore foot were plainly visible, whilst the hind foot showed five.

We crossed the stream, and then saw that, on the snowy grass, these separate foot-impressions almost merged themselves into one. This looked as though it might have been made by a one-legged animal, too.

But Jerry carefully brushed away the snow, and then I saw that the four separate pad-marks were still there, but almost superimposed on each other.

"Stoat," said Jerry, "he's bin boundin' along puttin' his front feet very close together, and then bringin' his back uns close up on top o' 'em."

We followed the trail. It went straight across the field.

"A weasel wouldn't 'ave gone over t' top o' t' ground like this," said my friend, with his eyes on the trail. "He'd 'ave gone underground — usin' a mole or a mouse run."

When the stoat's track had almost reached the distant hedge it suddenly stopped. We could see the marks of the four prints where it had stood squarely on all fours. Then the trail turned off directly to the right and went down the field.

"That's rather funny," said Jerry. "I wonder what made him turn so suddenly. He seemed to be on some hunt or other, and a stoat don't easily turn aside."

But though we searched long we could not trace him. Rabbits had crossed and re-crossed the track.

"O' course ye know which way that rabbit's travellin'," said Jerry, kneeling in the snow.

"You can't catch me this time, Jerry," I said, laughing.

"Lots o' folk think 'e's travellin' in t' opposite direction cos 'e brings them hind feet o' 'is in front o' fore uns when runnin'."

"It's a wonder to me that with all the enemies that a rabbit has, it still manages not only to live but also to thrive — and it can't be said to have much defence, can it?"

"Well," said Jerry, "it hasn't much of an attack certainly, and it doesn't use to the full what it has. Have ye ever bin ripped wi' a rabbit's hind claws?"

I shook my head. "Well," he continued, "allus pick up a wounded un carefully, fer with those high-power back legs he can slash right into yer arms and mak' a fearful mess o' 'em."

"I was seated on a bank like this," he continued, "and I saw a stoat comin' towards me, and, curious enough, a rabbit were follerin' it." He looked at me to see whether the stating of such a fact produced the astonishment which it ought to have done.

"I've seen dozens of rabbits chased by stoats. This un were fleein' from t' rabbit. Every now and then 'bunny' would rush up to him, tak' a flying leap right ovver him, and t' stoat ud lie quite close to t' ground. Once I'm certain she caught him a crack on his head which must 'ave made him pretty dizzy, fer he turned aside and dashed inter some stones. Then the rabbit went back to her young uns."

"She had a nest, then, somewhere or other?" I asked.

Jerry nodded. "I could see that she were a mother, fer she'd plucked all the soft fur from off her breast to mak' the young uns a snug bed. But it were a plucky thing fer 'er to do.

"And what did you do?" I asked.

"Oh, I waited aboot fer a few minutes, quiet-like, fer I half-expected t' stoat 'ud poke his head oot and 'chitter' like a monkey at me — tell me what he thowt aboot my comin' on t' scene. But I reckon his head must 'ave bin full o' that kick the rabbit give 'im fer I never saw 'im no more."

CHAPTER
TWENTY-TWO

The Little Fellow in Velvet

"I've a bit of a job up near t' Hall this mornin'. Would ye care fer a walk?" asked the gamekeeper as we reached his cottage door. "There's a fox's hole I want to fill. T'huntsmen are comin' round to-morrer. Would ye mind bringin' t' spade, as I've got me gun?"

So over the fields we tramped.

In the distance were the great cliffs on the fells.

"Git yer glasses out and fix them on that bit o' white sittin' on top o' yon black rock."

I did so, and then handed them to John.

"Aye," said he, "I fancied yon were a peregrine's front."

He handed me back the binoculars, and I saw that the falcon was sitting as still as death, with his head hunched into his shoulders and his round eyes staring at the valley beneath. No movement in the air or on the ground was being missed by those two cruel, keen, lustrous eyes.

We walked slowly, knowing full well that the bird had seen us, but hoping to get a nearer view.

Then John seized my arm and pointed to a pigeon that was flying at right angles to the peregrine's look-out. We both sank amongst the heather, and I felt Raq come and creep up against me. He knew not what was happening, but was shivering with the excitement we had communicated to him.

"We'll see summat worth seeing in a minute, or I'm a Dutchman," said John.

No sooner had he spoken than the falcon launched himself into the air with a swaggering roll. The pigeon saw his enemy at once, and his wing-beats quickened perceptibly. I could see the flash of the white bars on his wings, and he seemed to be stretching out his head more eagerly. At the same moment he changed his direction.

"He's goin' to make fer that clump o' trees," said John, pointing to a wood which lay some distance away in the valley. "It's his only chance."

How fast the two travelled! The morning light shone on the light blue back of the pigeon. The peregrine hurtled through the air like a dark thunderbolt.

"The pigeon's gaining a little," I said hopefully.

"Only fer a minute," said John. "Ye see 'ow t' peregrine is risin'. Watch 'im, he'll swoop on't like greased lightnin'. There!"

It was only too true. The falcon, after a fearful burst of speed, closed his wings and lowered his

157

head. Like an arrow of doom he fell, and his talons slashed into his victim's back. We heard the body crash on the ground.

"The savage brute," I said angrily.

"Aye, it's a cruel business," said John as he left me to go along the drive leading up to the Hall.

"I won't be above a few minutes if ye'll wait."

The time passed very quickly. Everything, trees, hedges, woods, fields, seemed to be in the lap of December peace.

Apart from the lazy cawing of the rooks, and an occasional querulous complaint of a peewit, the only live things were the large flocks of finches which rose and fell on the stubble.

"Come along, Raq," I called, as John appeared at the lodge gates, but Raq was not obedient. He seemed very intent on something.

He was looking at the ground, and for a moment he was so absorbed that he forgot to wag his tail.

When we reached the spot the keeper handed me his gun quickly, and seizing the spade, he drove it deep into the ground just where the surface appeared to be rising. Raq never moved. Then John gave the spade a dexterous twist, and the next moment he had brought to view a mole. I got the dog by his collar; otherwise I am afraid he would have made short work of the burrower.

Equally dexterously, John nipped the mole by the scruff of what ought to have been his neck, and held him out. I had never noticed till then that the little

158

fellow has no proper neck. Head and body join without any connecting link.

"'E's not exactly what ye might call a beauty, is 'e? 'E's just a tube wi' a long snout an' what 'e thinks is a tail."

John pointed to both extremities. "Ye'll often find," he continued, "that where they uses a lot o' material fer this end" (here he pointed to his long, quizzing nose) "they don't have much left fer t'other end."

I laughed, for the mole's stumpy tail does look an absurdity. "He doesn't need much of a tail if he's always running about in dark corridors."

"True," said John, "an' 'e fits into that burrer like a cartridge fits into yon gun, so there's no need to steer or swerve. Did ye ever see sic a pair o' front feet?"

I examined them, whilst the mole showed every sign of temper. They were not placed where front feet ought to be. They grew out of the neck that wasn't there! Then, too, instead of growing downwards, as all self-respecting feet should do, they grew sideways, so that the white crinkled soles seemed to be ready to push one away sideways. Each foot had sharp claws, too.

John drew my attention to his nose, then to his ears, and to his small slits of eyes. I could only manage to see them by blowing on his velvet pile.

"'E smells 'is worm first, under t' groond, then 'e 'ears it, and then, mebbe, 'e gits a squint on't."

"And then how does he tackle it?" I asked.

"Ower forty sharp teeth mak's mince-meat on't. If ye'll 'old on to Raq I'll let t' li'l feller go."

He put the mole on the ground, and for a moment we saw its dark body before us. Then it simply took a "header" into the earth and was not.

"What do ye think o' that fer a vanishin' trick, eh? 'E knows 'ow to use them spades of 'is, I reckon."

"You don't kill them when you come across them. then?" I asked, as we neared our destination.

"Only when they mak' a mess o' t' lawns near t' Hall. Then they're a nuisance."

"Is that the only damage they do?" I asked.

"I allus reckon, generally speakin', they do more good than 'arm. They mak' 'underds o' burrers, and they 'elp to drain t' soil. I've knowed farmers say as 'ow they bring good soil up to t' surface. O' course, it gives them the trouble o' scatterin' t' molehills and levellin' up a bit, but t' good they do to t' land is worth it. Why a keeper dislikes 'em is 'cos t' weasels use their runs to keep oot o' sight. Ye'll often catch a weasel when ye set a mole-trap."

The keeper paused a moment and looked round at the coverts which surrounded us. "On t' other 'and, I reckon they do us many a good turn by shovin' up the soil, fer there's allus a lot o' insecks in a mole-hill, an' t' partridges know it, so it works both ways, ye might say."

"Well, here's to the li'l feller in velvet, John," I said.

John smiled and raised an imaginary cup too.

CHAPTER
TWENTY-THREE

The Fox's Brush

"Have you started a training stable?" I asked Jerry, as I met him leading a fine Clydesdale.

The old poacher laughed. "Owd Graham up at Whiterigg yonder, 'as got lumbago, and this hoss of his wanted shoein', so I've been to t' blacksmith's wi' it."

"It's a fine beast," I said, rubbing its muzzle.

"If ye'll go along wi' me," said Jerry, "I've nowt else to do as soon as I've put 'er in t' stable."

"It looks as though in a few years' time a horse will be a rare sight," I said as we walked along. "People will stare as they did when motors first appeared on the roads."

"It looks like it, more's the pity; though i' this part o' t' country farmers are givin' up their tractors and goin' back to 'em. To my way o' thinkin', there's nowt better'n a hoss, unless it's a dog," said he, looking at Raq. "They seem to understand ye."

At the mention of the word "dog" Raq leaped up at Jerry's fingers, giving them a tender nip.

"I allus find that where there's hosses workin' and wand'rin' in t' fields ye find more game aboot an' all. Ye see, a tractor is such a quick workin' thing. It turns ower acre on acre i' record time. But with hosses, only a small piece of land were turned up fresh every day, and this allus used to attract t' birds."

"You mean the food was rationed out better," I said.

Jerry nodded. "Ploughin' lasted fer weeks, and each day t' plough kept bringin' insecks an' worms to t' top, an' when t' hosses 'ad a meal in t' fields, they tossed their 'eads when feedin', and they'd scatter a bit o' corn oot o' their nose-bags fer t' birds."

He patted the glossy neck of the newly-shod horse.

"Besides, when a few hosses were kept on t' farm, it meant that oats had to be growed so as to keep 'em, didn't it? An' t' straw were needed fer their beddin', and where there's corn there's birds. But wi' them tractor things" — here he turned to me — "all ye need is a few petrol cans an' a drum o' oil, and them don't encourage the birds."

Under the shadow of a hedge we watched the antics of some rabbits which had been lured out by the sunshine. How quaintly they sat up and performed their toilets! Even as a cat sits on her haunches, and, after moistening her paws, draws them down over her ears to the tip of her nose, so the rabbits washed their faces. Some of them had clay or sand on their hind legs, and it was amusing to see them flick their hind legs in quick kicks to clean them.

"They are cleanly little beggars," I said to Jerry.

"Aye, they are; but they've some queer ways an' all. Have ye ever smelt a fox, or 'is hole?"

I made a wry face.

"Well, the strange thing is that ye sometimes find rabbits livin' in t' same holes as foxes."

"I thought only badgers did that."

"They do, o' course, but ye'll also find 'e 'as rabbit lodgers."

"Then the fox won't have far to go for his meals. He must find it rather handy being able to slip from his bedroom and just dash up a corridor for his breakfast. But why on earth don't the rabbits clear out?"

Jerry puzzled me still more by saying, "An' I've even know'd t' rabbits to stop when t' vixen had 'er cubs."

I said nothing, and by my expression Jerry saw that I thought he was now pulling my leg.

"Ye've overlooked summat," he said. "If ye were chasin' a cat doon a road, and it suddenly bolted through a hole in t' fence, ye couldn't follow it, could ye? Well, t' rabbits need smaller 'oles than t' fox, and so when 'e chases them, they just skip oot o' t' main corridor up one o' their smaller runs. I reckon that for quick movin' aboot underground they can leave owd Reynard standin'."

"But what makes them live with him in the first instance?"

Jerry grinned. "It ain't likely that they goes to be 'is guests. As t' fox prowls aboot 'e comes across a nice sized 'ole made by t' rabbits, and 'e just settles doon among 'em wi'oot askin' 'by yer leave,' or t' price o' t' rent, or owt."

"And what about when the cubs arrive? Don't they make things rather warm for the rabbits? They can follow up the smaller runs, can't they?"

"I reckon t' rabbits only stay so long as t' cubs are small. Then they clears oot, you bet. Them smaller runs are 'andy for t' cubs if a terrier gets into their main burrer. They can bolt up the small rabbit runs for safety."

Walking along by the fringe of a stream, the old poacher pointed to clear made padmarks in the soft clay.

"Them's made by t' feller we're talkin' aboot — t' feller with the smell, ye know."

"They are rather neat — like a cat's, aren't they? Only these show clawmarks."

"Aye, an' ye've got a good run on 'em 'ere," nodded Jerry, at the same time pointing to a dozen tracks clearly defined in the soft ground. "He's bin 'untin' pretty carefully by t' looks on 'em."

All the marks ran in a straight line, and might have been made by a one-legged animal.

"When ye see 'em like that," said Jerry, "ye can gen'rally reckon as 'ow 'e's watched 'is steps pretty carefully. Mebbe 'e's bin stalkin' a water 'en or a vole. Ye see, 'is hind legs fit exactly in the 'oles made by 'is front 'uns. Compared wi' t' fox, that dog o' your'n is a untidy walker."

For a moment or two I watched Raq carefully, and when he had walked over a small sandy bay I saw exactly what Jerry meant. Raq's hind feet did not fit into the padmarks left by his front ones. They were "out of register," as a printer would say.

"Have ye ever seen a fox goin' ower puddly ground?" asked Jerry.

I shook my head. "He is very proud of his tail, isn't he? He wouldn't want that to get wet."

"Aye, you bet 'e 'is. But ye mustn't let a 'untsman 'ear ye call it a tail — it's 'is 'brush.' Them 'untin' fellers are rather perticlar aboot the way they talk aboot owd Redcoat. T' foot is his 'pad' an' 'is 'ead is his 'mask,' an' a fox-hound is never a dog, allus a ''ound,' mind ye. But I was goin' to tell ye aboot 'ow 'e goes ower puddly ground. Ye're right when ye say that 'is brush is 'is pride. When 'e 'breaks' from t' wood wi' the 'ounds after 'im, it's a fine sight to see 'im flourish 'is brush, defying 'em. A fox that's got a bit o' spunk in 'im gives it a rare whisk, as though he's tellin' 'em to do their d —" Jerry pulled up just in time, and spluttered out rather lamely, "dirtiest."

"That was a weak one, Jerry. You only just managed to clear that fence," I said, with mock severity.

The old sinner grinned.

"I'm still waiting to know how that fox walks over puddly ground."

He laughed. "I seem to have lost t' scent a bit. Howsomever, I'll 'break back' to it. He walks ower puddles wi' 'is tail up as stiff as 'e can git it. 'E'd lose 'is dignity if 'e was to trail it in t' mire. A collie dog doesn't lift his tail up, though a hoss, when he's roused and comes after ye, curves 'is. Some day, when we're oot together, we'll foller t' hunt, and if ye see t' fox wi' its 'brush' up, then ye'll know 'e's still plenty o' life left in 'im. But if ye see the 'ounds after 'im an' 'is tail lyin' limp behind 'im, ye'll need to feel sorry fer 'im. That's the beginnin' o' t' end."

"Poor old Reynard," said I.

For a moment or two the scene of the hunted fox nearing its end passed before my eyes. That drooping tail would speak of waning courage and spent strength. Bedraggled, too, it would be. No thought of pride, but only fear of death would be uppermost — "a pack wi' t' devil in 'em at 'is 'eels," as Jerry had vividly put it.

I think I must have sighed audibly, for Jerry looked sharply at me and asked —

"What's up?"

"I don't think I'll follow the hounds with you, Jerry," I said.

For answer he linked up his arm in mine, and Raq looked up at us as though he, too, understood.

166

CRAKE!

"OLE BILL"

WATER VOLE

Bm   Whp.
Ball   HP.
HPe
GP
Sw
R Pl
W P
MM
Wat,
Brh.